机械制造自动化系统

（双语版）

杨贵超　[爱尔兰]露西 · 麦考利（Lucy McAuley）　王　华　主编

Mechanical Manufacturing Automation System

·北京·

内容简介

本书系统介绍了机械制造自动化领域的基本概念、基本理论和基本方法，适当反映机械制造自动化方面的最新进展，内容包括：工业自动化、气动学、可编程逻辑控制器、逻辑系统和PLC编程、顺序控制、PID控制、机器人、视觉系统、数控技术。

本书适合高等院校中外合作办学的本科生（机械类）、留学生（机械类）以及机械类和自动化类相关专业本科师生学习使用，也可供从事机械制造自动化等工作的工程技术人员使用。

图书在版编目（CIP）数据

机械制造自动化系统：双语版：英汉对照/杨贵超，（爱尔兰）露西·麦考利（Lucy McAuley），王华主编. —北京：化学工业出版社，2021.2（2023.8重印）
ISBN 978-7-122-38131-6

Ⅰ.①机… Ⅱ.①杨…②露…③王… Ⅲ.①机械制造-自动化技术-高等学校-教材-英、汉 Ⅳ.①TH164

中国版本图书馆CIP数据核字（2020）第243471号

责任编辑：丁文璇　　装帧设计：张　辉
责任校对：刘　颖

出版发行：化学工业出版社（北京市东城区青年湖南街13号　邮政编码100011）
印　　装：北京七彩京通数码快印有限公司
787mm×1092mm　1/16　印张11　字数264千字　2023年8月北京第1版第3次印刷

购书咨询：010-64518888　　售后服务：010-64518899
网　　址：http://www.cip.com.cn
凡购买本书，如有缺损质量问题，本社销售中心负责调换。

定　　价：48.00元

前言

Preface

为了适应新时代高等院校机械类专业本科生源面向国际化的需求，在充分吸收相关学科发展的新理念、新知识、新方法、新手段的基础上，我们编写了本书，力求在内容和形式上体现机械制造自动化类人才培养的要求与特点。

本书主要面向高等院校中外合作办学的本科生（机械类）、留学生（机械类）以及机械类和自动化类专业方向的本科生，不但可为学生奠定坚实的机械、自动化方面的知识基础，而且有利于提高学生的英文理解和应用能力，为培养国际型人才打下良好基础。

本书共9章。第1章概述工业自动化系统；第2章介绍气动学方面的基础知识；第3～5章分别介绍可编程逻辑控制器（PLC）、逻辑系统和PLC编程、顺序控制方面的内容；第6章着重介绍PID控制；第7章介绍机器人方面的基本知识；第8章介绍视觉系统方面的知识；第9章介绍数控技术。本书系统性地介绍机械和自动化方面的基本概念、基本理论和实现自动化的基本方法，内容浅显、易于理解，其主要特点包括：

（1）内容全面。本书内容不但涵盖气动学、电气学、可编程逻辑控制、机器人、视觉系统等常用的机械制造自动化知识，还重点介绍了工业中常用的PID控制、图像处理技术、数控技术等，内容组织注重基础性、系统性，同时兼顾应用性、先进性。

（2）难度适中。本书内容以知识介绍为主，绝大多数知识点没有深入展开，难度不大。同时贯彻少而精、理论联系实际、学以致用的原则，在介绍基本知识的基础上，引入常见的应用案例，如气动阀控制气缸运动、工业系统可编程逻辑控制、机器人运动控制等。

（3）编排合理。教材采用双栏排版，左侧英文，右侧中文，有助于学生脱离中文环境而掌握自动化知识领域的英文表达。中文部分也非完全翻译英文，重在指导学生理解英文内容。插图生动而不失严谨，在保证自然科学严肃性的同时，有利于激发学生的阅读兴趣。

本书的第1～4章、第6章由杨贵超、Lucy McAuley、王华编写；第5章和第9章由Lucy McAuley、方成刚、张浩、孙付仲编写；第7章和第8章由Lucy McAuley、张浩、孙付仲、孟龙晖编写。全书由杨贵超、Lucy McAuley、王华担任主编并统稿。同时，感谢Paul Dillon给本书提供素材。

由于编者水平有限，书中难免有不妥之处，恳请广大读者批评指正。

编者

2020年10月

目录

Contents

Chapter 8

Vision System / 111

第 8 章

视觉系统 / 111

Chapter 9

Numerical Control Technique / 128

第 9 章

数控技术 / 128

Chapter 1 Introduction of Industrial Automation
第1章 工业自动化简介

Key words 重点词汇

automated systems 自动化系统
component 部件，组成部分
circuit design 电路设计
pneumatics 气动学
programmable logic controller 可编程逻辑控制器
electropneumatics 电气学
robotics 机器人学
machine vision 机器视觉
machine vision 机器视觉

With the continuous progress of science and technology, the machinery manufacturing industry has entered the era of automation. As the main component of manufacturing automation, mechanical manufacturing automation is the basis for enterprises to realize automated production and participate in market competition. This chapter briefly introduces pneumatics and electropneumatics, PLC, robotics, machine vision and other contents in industrial automation.

随着科学技术的不断进步，机械制造业已进入自动化的时代。作为制造业自动化的主要组成部分，机械制造自动化是企业实现自动化生产、参与市场竞争的基础。本章简单介绍工业自动化中气动学与电气学、PLC、机器人学、机器视觉等内容。

1.1 Pneumatics and Electropneumatics

Electropneumatic systems are widely used in many areas of industrial automation, for example in production, assembly, and packaging systems worldwide. These systems are driven by electropneumatic control systems and are used to move, position and clamp components. In electropneumatics, pneumatic components are controlled by circuit. Normally, electronic and electromagnetic sensors, electrical switches and programmable controllers are used to replace the manual control of a pneumatic system. This is covered in detail in Chapter 2.

1.1 气动学和电气学

电气系统广泛应用于工业自动化中的众多领域，例如遍及全世界的生产、装配和包装系统。这些系统由电气控制系统驱动，用于移动、定位和夹紧部件。在电气学中，气动元件由电路和电子电路来控制。通常，用由电子和电磁传感器、电气开关和可编程控制器等组成的控制系统来代替气动系统的手动控制。详细内容参见第2章。

1. 2 Programmable Logic Controller

1. 2 可编程逻辑控制器

A programmable logic controller (PLC) uses a software program to read information from the input terminals and control the outputs. PLC can monitor and control a large number of inputs and outputs, so that even very large processes can be controlled. The control sequence performed by the PLC is determined by the program loaded into it. Each PLC can have a different program, and the program in a particular PLC can be subsequently altered, all of which makes for very flexible control. This is covered in detail in Chapter 3～Chapter 5.

可编程逻辑控制器（PLC）使用软件程序从输入端子读取信息并控制输出。PLC 可以监视和控制大量的输入和输出，即使非常大的过程也能控制。PLC 可执行的控制序列取决于加载到其中的程序。每个 PLC 可以有不同的程序，特定 PLC 中的程序可以被更改，所有这些都使控制变得非常灵活。详细内容参见第 3～第 5 章。

1. 3 Robotics

1. 3 机器人学

Robotics has been greatly developed at present. Fig. 1-1 shows the data from the International Federation of Robotics (IFR) report 2018 and shows the increase in the estimated number of robots in use world-wide since 2009.

目前，机器人技术得到了巨大的发展。图 1-1 为国际机器人学联合会（IFR）2018 年度的报道数据，该数据反映了自 2009 年以来全球在用机器人的估计数量，其一直处于增长趋势。

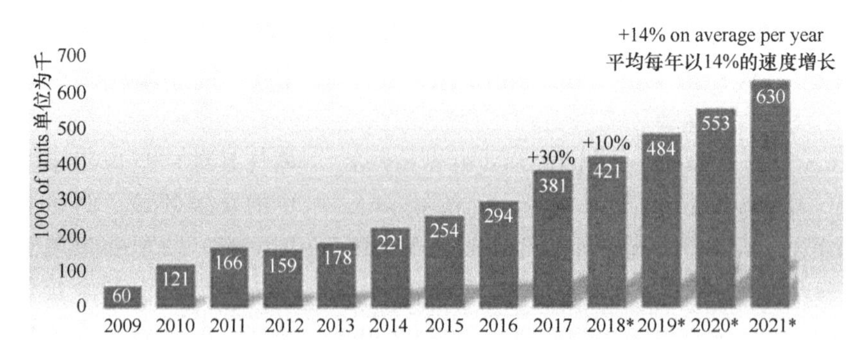

Fig. 1-1 Worldwide supply of industrial robots

图 1-1 全球工业机器人的供应情况

This growth is particularly true in China which purchased the largest number of robots in 2017, as shown in Fig. 1-2. Fig. 1-3 shows the main application areas for industrial robots and the percentage increase

如图 1-2 所示，2017 年购买机器人数量最多的国家是中国，由此可见，对机器人技术的需求在中国尤为明显。图 1-3 显示了工业机

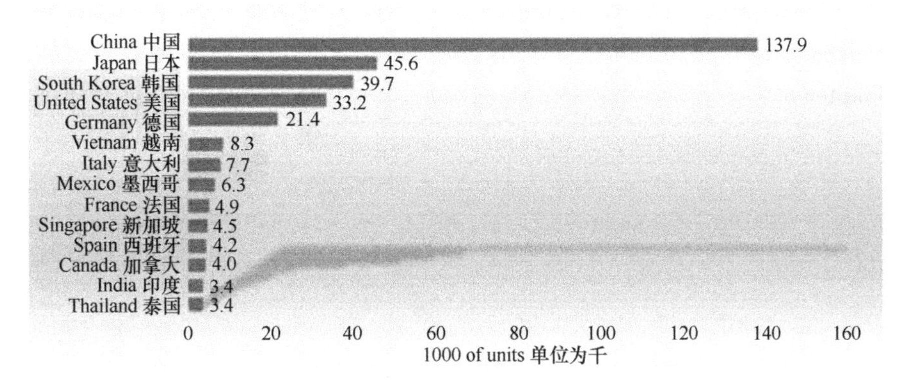

Fig. 1-2　Estimated number of robots worldwide

图 1-2　全球工业机器人的估计数量

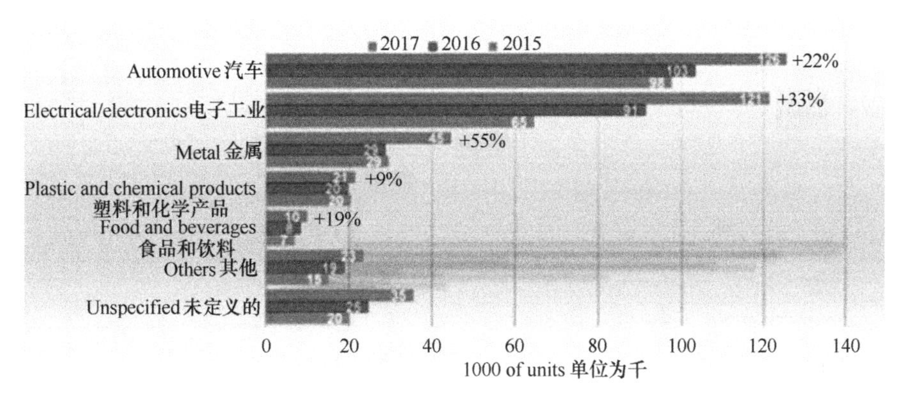

Fig. 1-3　Application areas of industrial robots

图 1-3　工业机器人的应用领域

from 2015 to 2017. The key components of industrial robots and the programming language V+ will be introduced in Chapter 7.

器人的主要应用领域及其从 2015 年到 2017 年的增长率。工业机器人的关键部件和编程语言 V+将在第 7 章进行介绍。

1.4　Machine Vision

1.4　机器视觉

Machine vision is used widely in inspection and control of automated systems. A camera takes an image which is then analysed using image processing software. The output data can be used to make a decision about a part, e. g. , is it present, is it the correct part, is it correctly

机器视觉广泛应用于自动化系统的检查和控制。图像处理软件对照相机拍摄的图像进行分析，其分析后的输出数据可用来对零件做出决策，比如，零件是否存在，

positioned, does it have the correct dimensions? The data can also be used to monitor the overall process and record information such as number of parts, number of errors and type of errors. This is covered in detail in Chapter 8.

是否是合适的零件，是否定位准确，是否具有准确的尺寸等。这些数据还可用于监控整个过程并记录诸如零件数量、错误数量和错误类型等信息。详细内容参见第8章。

Chapter 2　Pneumatics
第2章　气动学

Key words 重点词汇

compressed air　压缩空气
atmospheric pressure　大气压
compressor　压缩机
actuator　执行器
cylinder　气缸
single-acting cylinder　单作用缸
double-acting cylinder　双作用缸
piston rod　活塞杆
directional control valve　方向控制阀
flow control valve　流量控制阀
air service unit　气源装置
poppet　锥阀
5/2 way valve　二位五通阀
pneumatic circuit diagram　气动回路图
solenoid　螺线管
switch contact　开关触点
normally closed　常闭
normally open　常开
limit switch　限位开关
reed switch　簧片开关
wiring of sensors　传感器接线
relay　继电器
latch circuit　锁存电路

2.1　Pneumatics Overview

Pneumatic transmission uses compressed air to work. Compressed Air is defined as air at a pressure greater than atmospheric pressure (1 atm). When using compressed air to operate a system, the pressure and the flow rate of the air need to be controlled.

Pressure is defined as the force per unit area. Currently there are various units used to characterize pressure, such as bar, Newton per square metre (N/m^2) and pounds per square inch (psi). N/m^2 is the SI (international standard) unit, bar is a metric unit but do not belong to SI system, psi is an imperial unit. The conversions between them are as follows:

2.1　气动学概述

气压传动利用压缩空气来进行工作。压缩空气是指压力大于大气压力（1 atm）的空气。当系统采用压缩空气作为传动介质来运行时，需要控制空气的压力和流速。

压力定义为单位面积上的作用力。目前，有多种用于表征压力的单位，如巴（bar）、牛顿每平方米（N/m^2）和磅每平方英寸（psi）。其中，N/m^2 是压力的 SI（国际标准）单位，bar 是公制单位但不属于 SI 系统，psi 是一种英制测量单位。它们之间的换算关系如下：

$$1\text{Pa}=1\text{N/m}^2$$
$$1\text{bar}=1\times10^5\text{N/m}^2=100\text{kPa}$$
$$1\text{mbar}=1\times10^{-3}\text{bar}$$
$$1\text{psi}=68.95\text{mbar}$$

The normal pressure range used in industrial applications is 4～10 bar.

工业应用中常用的压力范围为 4～10 bar。

The frequently-used flow units in the pneumatic industry are litres (L), cubic decimetres per second (or dm^3/s) or cubic meters per minute (m^3/min).

气动工业中常用的流量单位为：升（L）、立方分米每秒（dm^3/s）或立方米每分钟（m^3/min）。

2.2 Production of Compressed Air

2.2 压缩空气的产生

Pneumatic control systems depend on the supply of compressed air, which must have sufficient quantity and certain pressure to satisfy the capacity of the system.

气动控制系统依靠压缩空气的供给来运行，这意味着必须提供足量并且具有一定压力的压缩空气，才能满足系统的功率。

A compressor is a machine that compresses the air or another type of gas from a low inlet pressure (usually atmospheric pressure) to a higher desired pressure. A compressor increases the pressure of gas by reducing its volume. Work required for increasing pressure of gas is available from the prime mover driving the compressor. Generally, the prime mover may be an electric motor, internal combustion engine, steam engine or turbine. Compressors are classified based on their operating principle, namely positive displacement compressors and dynamic compressors, as shown in Fig. 2-1.

压缩机是一种将空气或其他类型的气体从低入口压力（通常指大气压力）压缩到更高的期望压力的机器。压缩机通过减小其容积来增加气体的压力。利用原动机驱动压缩机，可实现气体的加压过程。一般情况下，原动机可以是电动机、内燃机、蒸汽机或涡轮机。根据压缩机的运行原理进行分类，可以分为容积型压缩机和速度型压缩机，如图 2-1 所示。

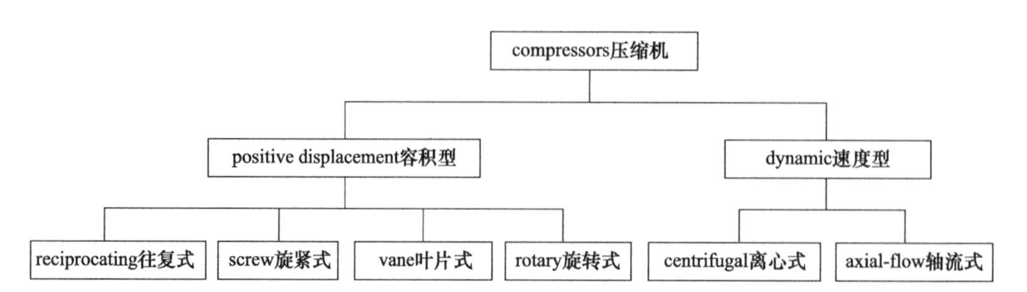

Fig. 2-1 Types of compressors

图 2-1 压缩机的类型

In positive displacement compressor, the compression function is acheived by shitting the solid boundary, which prevent the flow from flowing back along the pressure gradient. Due to solid wall displacement these compressors can produce quite large pressure ratios. Positive displacement compressors are further classified based on the type of mechanism used for compression, namely reciprocating compressors, rotary compressors, screw compressions and vane compressions.

在容积型压缩机中，通过固体边界的移位来实现压缩功能，固体边界能够阻止流体沿压力梯度方向回流。由于固体壁的移位，这些压缩机能够产生相当大的压力比。根据实现压缩功能的机构类型可将容积型压缩机进一步分为往复式压缩机、旋转式压缩机、旋紧式压缩机和叶片式压缩机。

Dynamic compressors, also called steady flow compressors, its performances depend on the momentum and inertia of a fluid. Rather than physically reducing the volume of a captured pocket of gas, dynamic compressors instead speed up the gas to high velocity, and then restrict the gas flow so that the reduction in velocity causes pressure to increase. Dynamic compressors may be axial-flow type or centrifugal type dependi ng on the type of flow in the compressor.

速度型压缩机也称为定常流压缩机，其性能取决于流体的动量和惯性。速度型压缩机并不是物理地减小容腔的体积，而是将气体速度提高，然后限制其流动，使得气体的速度降低从而压力增加。根据压缩机中流量的类型，速度型压缩机可分为轴流式或离心式。

2.3 Components of a Pneumatic System

2.3 气动系统的组成

Fig. 2-2 shows the main components of a pneumatic system, including a cylinder, air line, flow control valve and air service unit.

图 2-2 列出了一个气动系统的主要组件，包括气缸、空气管路、流量控制阀和气源装置等。

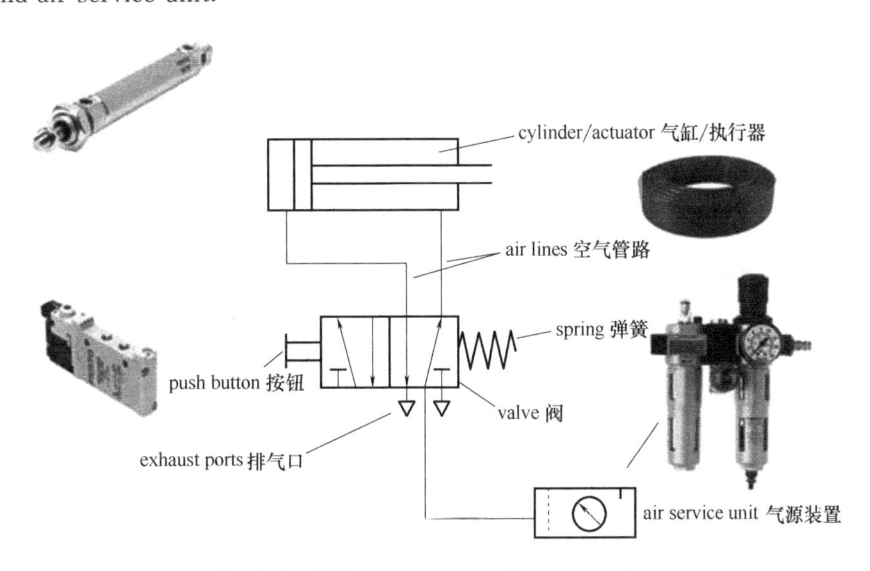

Fig. 2-2 Main components of a pneumatic system
图 2-2 气动系统的主要组件

2. 3. 1 Air Service Unit

Compressed air is an essential power source that is widely used throughout industry. However, compressed air may contain water and particles of dirt and wear which all mix together to form an unwanted condensate. This condensate is often acidic and can cause wear in tools and pneumatic equipment, block valves and corrode piping which leads to high maintenance and costly air leaks. An air service unit, such as that shown in Fig. 2-3, is used on each piece of automated equipment to ensure that the compressed air supply is clean and dry, the pressure is at the correct level and fine particles of oil are drawn into the air to lubricate the parts within the system.

2. 3. 1 气源装置

压缩空气是在整个工业中被广泛使用的重要动力源。然而，压缩空气中可能会含有水分、灰尘、磨损颗粒，若它们混合在一起，将会形成多余的凝结物。这种凝结物通常呈现出酸性，会加速器具和气动机械的磨损，阻塞阀门和节流孔，从而带来很高的维护成本，造成严重的空气污染。通常，每套自动化设备都会利用如图 2-3 所示的气源装置，以确保供应清洁、干燥且压力稳定的压缩空气，并将雾化的润滑油混入压缩空气以润滑阀、气缸等元件内容易磨损的部分。

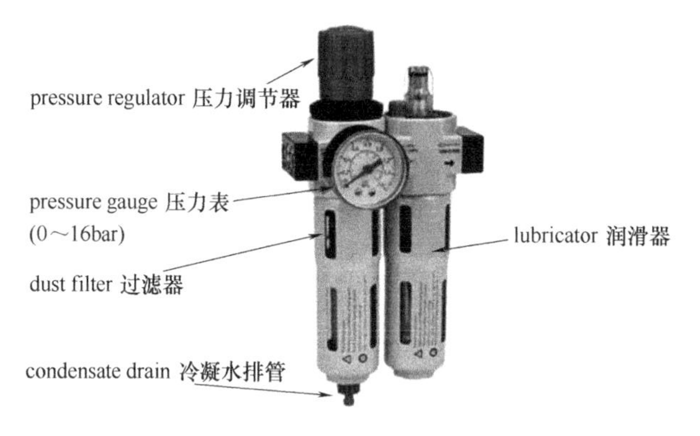

Fig. 2-3 Air service unit

图 2-3 气源装置

2. 3. 2 Pneumatic Actuator

Pneumatic actuators are mechanical devices which use the power of compressed air to produce a force in a reciprocating linear or rotary motion. The piston of pneumatic actuator is a disc or cylinder, and the piston rod transfers the force to the object to be moved. Examples of linear and rotary cylinders can be found in Fig. 2-4.

2. 3. 2 气动执行器

气动执行器是将压缩空气的压力能转换为直线往复运动或旋转运动等机械能的装置。气动执行元件的活塞为圆盘或圆柱体，活塞杆将其产生的力量传递给物体使其运动。其中，线性和旋转气缸的示例见图 2-4。

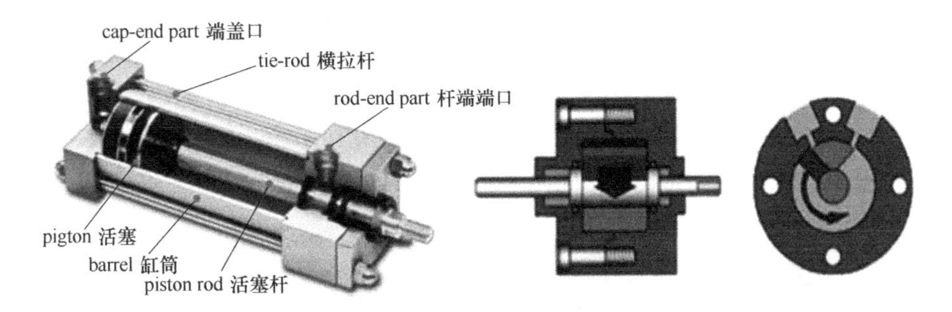

Fig. 2-4　Examples of linear and rotary cylinder

图 2-4　线性和旋转气缸示例

The inside diameter (bore) of the cylinder determines the maximum force and the stroke determines the maximum linear movement, as shown in Fig. 2-5. The maximum working pressure depends on the design of the cylinder.

气缸的内径（口径）决定其最大输出力，行程决定其最大线性运动距离，如图 2-5 所示。另外，最大工作压力取决于气缸的设计。

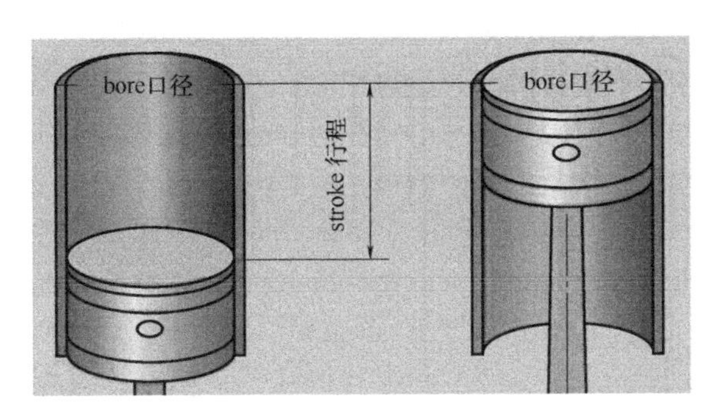

Fig. 2-5　Bore and stroke of a pneumatic cylinders

图 2-5　气缸的口径和行程

Single-acting Cylinders (Fig. 2-6)　Use compressed air to create a driving force in one direction (usually out), and a spring to return to the retracted position. This type of cylinder can have limited extension due to the space the compressed spring takes up. Another downside is that some of the force produced by the cylinder is used to compress the spring.

单作用气缸（图 2-6）　利用压缩空气在一个方向（通常向外）上产生驱动力，并通过弹簧来复位。由于压缩弹簧会占据一定的空间从而使这种类型的气缸使用范围有限。另一个缺点是气缸产生的部分力会因弹簧的作用而损失。

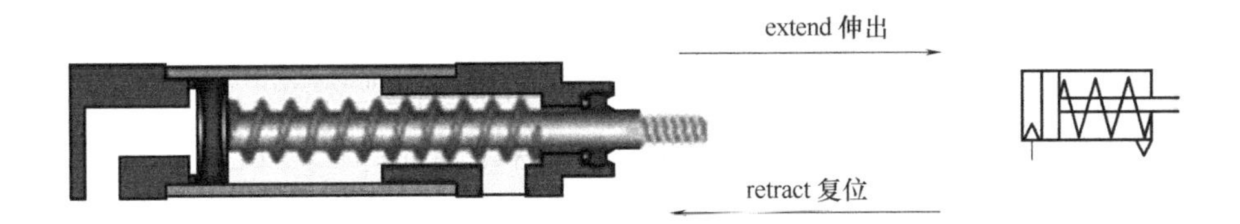

Fig. 2-6　Single acting cylinder

图 2-6　单作用气缸

Double-acting Cylinders (Fig. 2-7) Use compressed air to both extend and retract. They have two ports to allow air in, one for extension and one for retraction.

双作用气缸（图 2-7） 利用压缩空气进行伸出和缩回。它有两个端口用于空气的流入，其中一个用于实现伸出，另一个用于实现缩回。

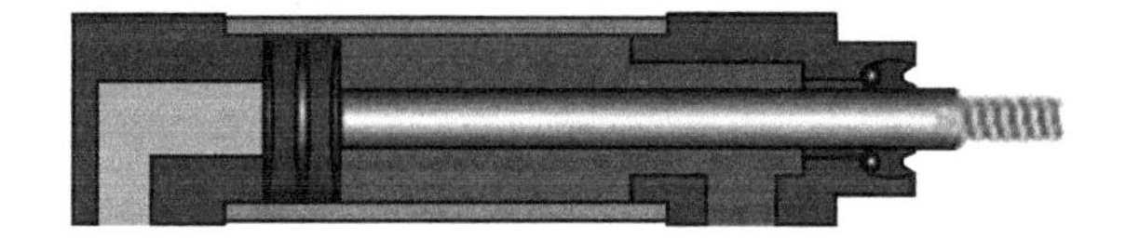

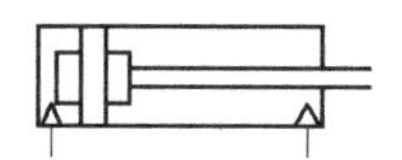

Fig. 2-7 Double acting cylinder

图 2-7 双作用气缸

2.3.3 Directional Control Valve

2.3.3 方向控制阀

Valves control the flow of air to the actuators and as a result control the force, velocity and direction of movement of the pneumatic actuator. Fig. 2-8 shows how the flow of air to a pneumatic actuator is controlled. The valve is a 5/2 way valve. This means that it has five ports (which are the openings in the valves) and two possible positions. As shown in Fig. 2-8, the air supply is connected to port 1. Port 4 and Port 2 are connected to either side of the pneumatic cylinder. Port 5 and Port 3 are exhausts where the compressed air is released to the atmosphere.

阀能够控制流进执行器的空气流量，进而控制气动执行器的力、速度和运动方向。图 2-8 展示了如何控制流进气动执行器的空气。该阀是一个 5/2（二位五通）阀，这意味着它有五个端口（阀上的开口）和两个工作位置。如图 2-8 所示，气源连接到端口 1，端口 4 和端口 2 连接到气缸的任一侧，端口 5 和端口 3 是用于将压缩空气排放到大气中的排气口。

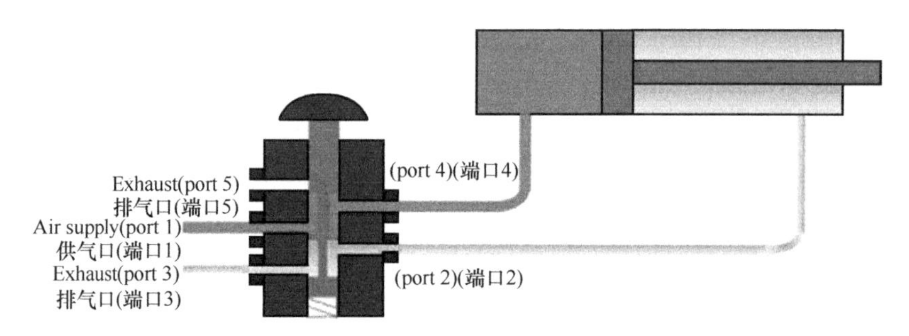

Fig. 2-8 Control of air flow to actuator

图 2-8 控制气流流向执行器

A push button valve is shown in Fig. 2-8. When the push button is pressed the valve will be in the position. Compressed air enters the valve through port 1 and

图 2-8 左边是一个按钮阀。当按下按钮时，阀芯将处于图示位置，压缩空气通过端口 1 进入阀并流向

is directed to port 4 which will cause the cylinder to extend. The air from the other side of the piston is released into port 2 and through port 3 into the atmosphere. When the push button is released the spring will push the spool in the valve back up. This will block exhaust port 3, air will flow from port 1 to port 2 and from port 4 to port 5. This will cause the cylinder to retract.

端口 4，从而使气缸伸出；同时活塞另一侧的空气流进端口 2 并通过端口 3 排入大气。当释放按钮时，弹簧将推动阀芯向上运动，排气口 3 被阻塞，空气将从端口 1 流向端口 2；同时活塞另一侧的空气从端口 4 流向端口 5，从而使气缸缩回。

The most common pneumatic valves are 3/2 way valve and 5/2 way valve. Fig. 2-9 shows the internal structure and symbol of a 3/2 way poppet valve. In the normal position the compressed air flowing into port 1 helps the spring to hold the poppet shut. Port 2 is connected through the plunger to the exhaust port 3. When the valve is operated the poppet valve moves down, the exhaust port is sealed and air flows from port 1 to port 2.

最常见的气动阀为 3/2（二位三通）阀和 5/2 阀。图 2-9 显示了 3/2 锥阀的内部结构和符号。在正常位置，锥阀在弹簧和流入端口 1 的压缩空气的同时作用下保持关闭。端口 2 通过柱塞与排气口 3 连接。当阀处于运行状态时，锥阀向下运动，排气口被阻塞，空气从端口 1 流向端口 2。

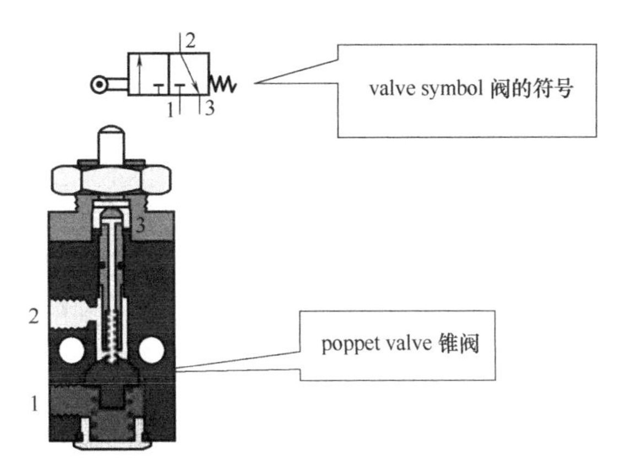

Fig. 2-9 Internal structure and symbol for 3/2 way poppet valve
图 2-9 3/2 锥阀的内部结构和符号

Fig. 2-10 shows the internal structure of a 5/2 way spool valve. In the normal position (when side 12 is activated) port 1 is joined to port 2 and port 4 is joined to port 5 (exhaust). In the operated position (side 14) port 1 is joined to port 4 and port 2 is joined to port 3 (exhaust).

图 2-10 为 5/2 阀的内部结构。在正常位置（12 这一侧被激活时），端口 1 与端口 2 连通，端口 4 与端口 5（排气口）连通。在工作位置（14 这一侧），端口 1 与端口 4 连通，端口 2 与端口 3（排气口）连通。

This type of valve is called memory valve as it remembers the last position to which it was switched even

这种类型的阀被称为记忆阀，它能够保持最终的切换位置，即使在

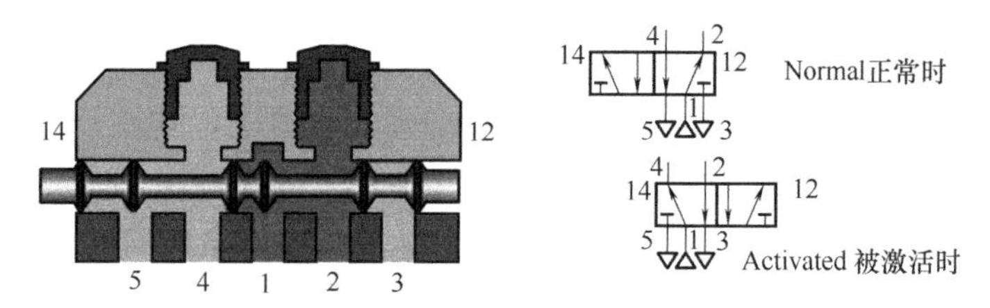

Fig. 2-10 Internal structure and symbol for 5/2 way valve

图 2-10 5/2 阀的内部结构和符号

when the signal to switch is removed. The air flowing around the spool holds it in position until a force is applied in the opposite direction.

切换信号消失的情况下也是如此。绕着阀芯流动的空气能够保证阀芯处于当前位置，直到在相反方向上施加作用力。

2.3.4 Flow Control Valve

2.3.4 流量控制阀

A flow control valve is used to control the speed of the cylinder as it extends and retracts. As shown in Fig. 2-11, the gas is allowed to flow into the valve without restriction but as the gas leaves the cylinder the one-way valve closes and the gas is forced to flow through the throttle valve. The gas cannot leave the cylinder at full speed so the movement of the cylinder is slower. The amount of throttle valve can be adjusted. Gas is always restricted as it leaves the cylinder (never as it goes into the cylinder) as it would otherwise affect the pressure of the gas supply to the cylinder.

流量控制阀用于控制气缸伸出和缩回时的速度。如图 2-11 所示，当气体通过流量控制阀进入气缸时不受约束，但是当气体从气缸流出到流量控制阀时，单向阀就会关闭，此时气体被迫流过节流阀，从而使气体不能全速地流出气缸，因此执行器的运动速度较慢。节流阀的开口是可以调整的。气体流出气缸时总是受到限制的（气体流入气缸时从不会受到限制），否则它会影响气缸内供应气体的压力。

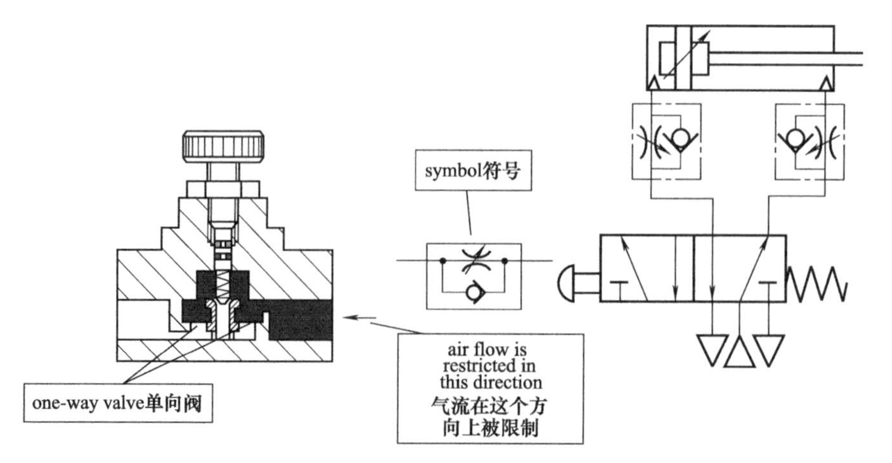

Fig. 2-11 Flow Control Valve

图 2-11 流量控制阀

2.3.5 Pneumatic Circuit Diagram

2.3.5 气动回路图

A pneumatic circuit diagram is used to indicate how the elements of the pneumatic system are to be connected together to give the correct movement of the actuator. Standard symbols are used for each element. A pneumatic circuit diagram is shown in Fig. 2-12. When the push button is pressed the cylinder extends and when the push button is released the cylinder retracts. Internal springs return the valve and the cylinder.

气动回路图用于表明如何将气动系统的元件连接在一起，以给出执行器的正确运动形式，其中每个元件都使用标准符号。图 2-12 所示为一气动回路图，当按下按钮时，气缸伸出；当松开按钮时，气缸缩回，内部弹簧也返回阀门和气缸。

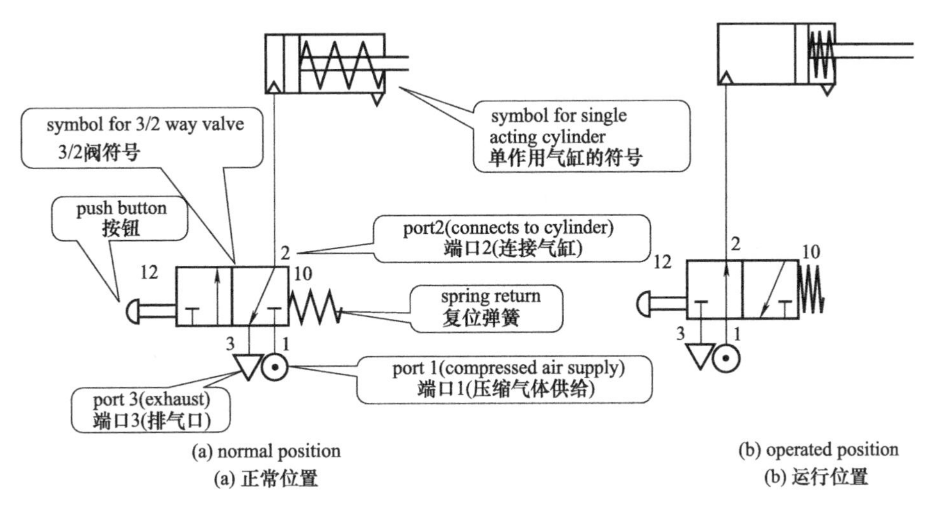

Fig. 2-12 Pneumatic circuit

图 2-12 气动回路

A three-way valve provides the inlet, outlet and exhaust path for the compressed air and is the normal choice for control of a single acting cylinder.

三通阀用于给压缩空气提供进气口、出气口和排气通道，通常用来控制单作用气缸。

In the normal position, i. e. , the button is not pushed, the valve is closed. This is how the circuit should be drawn, with the valve and the cylinder in the retracted position. Air is flowing from the cylinder into the exhaust port (port 2 to port 3). In the operated position, the valve is actuated by the push button. Air flows from port 1 to port 2 and into the cylinder. The cylinder extends. The push button must be held down to keep the cylinder extended. When the push button is released the cylinder retracts.

在正常位置，即未按下按钮时，阀处于关闭状态。此时，在画气动回路图时，阀和气缸处于缩回位置。空气从气缸流入到排气口(端口 2 到端口 3)。当通过按钮启动阀后，阀芯处于工作位置。空气从端口 1 流向端口 2 并进入气缸，气缸伸出。要想保持气缸一直处于伸出状态，就必须一直按住按钮。当按钮释放时，气缸会缩回。

2.4 Electropneumatic System Component

2.4 电气系统元件

In an electropneumatic system electrical signals from switches and sensors are used to control the valves. A number of basic components are used to constitute a electropneumatic system so before looking at the design of the system it is important to look at these elements, for example solenoids, switches, sensors and relays etc.

在电气系统中，通过开关和传感器发出的电信号来控制阀。由于电气系统由许多基本元件组成，因此在设计电气系统之前，首先要认识并了解这些元件，比如螺线管、开关、传感器和继电器等。

2.4.1 Solenoid

2.4.1 螺线管

In an electropneumatic system actuation of the valve is through the use of a solenoid, as shown in Fig. 2-13. A solenoid is a constructed by shaping a length of conductor into a number of loops (windings) that are packed closely together, leaving a construction like a hollow shaft. If a current is passed through the conductor, then a magnetic field is created in core of the windings. This will cause iron object placed within the core to move in the direction of the magnetic field. Therefore, the iron core can be attached to the actual spool in the directional control valve. To reduce the energy used the solenoid might operate a small ari valve which operates the main valve, called a pilot operated valve.

在电气系统中，通过使用螺线管来启动阀，如图 2-13 所示。将一段导体制造成许多紧密结合在一起的回路（线圈），就形成了螺线管，其结构就像空心轴。如果在导体中通入电流，线圈中则会产生磁场，从而使放置于线圈内的铁质物体沿磁场方向运动。因此，铁芯可以用来连接方向控制阀中的实际线轴。为了减少能耗，螺线管可以通过操作一个小的气阀来操作主阀，该气阀称为先导阀。

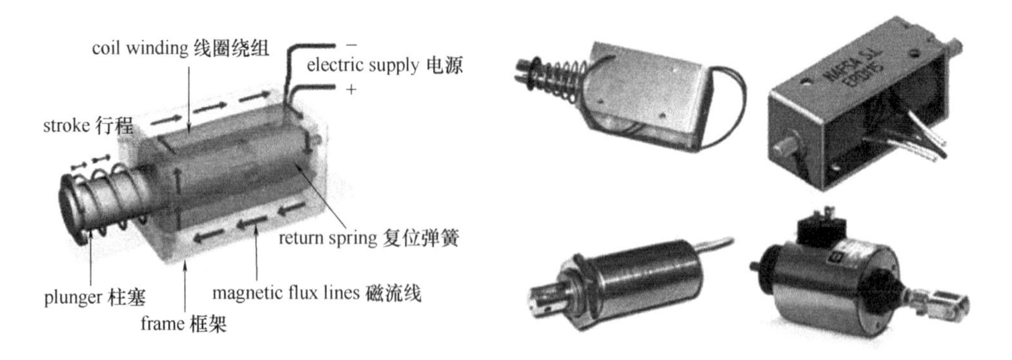

Fig. 2-13 Solenoid construction and typical industrial solenoids

图 2-13 螺线管结构和典型的工业螺线管

2.4.2 Switch

2.4.2.1 Switch Selection

The various types of switches with their associated design characteristics are so widely that it is impossible to go to any great detail but a few pointers on the electrical limits are useful.

The switch assembly is selected on the basis of the type of control it switches and is dependent on the operating voltage and on the operating current. Switches without any electronic circuitry attached have only a maximum voltage, maximum current and maximum power. Switches with electronic circuitry will have to draw some current in order to operate the electronics and so will have a minimum operating current.

2.4.2.2 Switch Position

A switch can be in one of two states, either on or off. This can be extended to take account of the normal condition of switch either closed or open in the "off" position. The conditions are described as "normally closed" or "normally open". The normally open (NO) switch does not allow electrical current to flow through its contacts under normal conditions while the normally closed (NC) switch does. Sometimes the contacts may be referred to as poles and will be numbered. The numbering system is shown in Fig. 2-14 and explained as follows:

- Normally open: path 3 to 4.
- Normally closed: path 1 to 2.
- Change over contacts: path 1 to 2 or path 1 to 4.

In the case of the linked contact the first pair of contacts is given the prefix 1 the subsequent 2, etc. along with 1, 2 to identify the path.

2.4.2 开关

2.4.2.1 开关的选择

开关的类型非常广泛，其设计特征又不尽相同，因此不可能掌握其所有重要的细节，但是掌握电气使用范围方面的一些要点是非常有用的。

通常开关组件是根据控制类型以及工作电压和工作电流来选择的。不含任何电子电路的开关只有最大电压、最大电流和最大功率。含有电子电路的开关需要一定的电流才能使电子元件工作，因此其具有最小的工作电流。

2.4.2.2 开关位置

开关可以处于开启或关闭两种状态，进而可以被扩展，考虑开关在“关闭”位置时的正常状态是闭合还是断开的。这些状态称为“常闭”或“常开”。常开（NO）开关在正常状态下不允许电流流过其触点，而常闭（NC）开关却允许。有时，触点也会被称为极点并被编号。图 2-14 所示的编号系统，其含义解释如下：

- 常开：路径 3 到 4。
- 常闭：路径 1 至 2。
- 转换触点：路径 1 到 2 或路径 1 到 4。

在转换触点的第一对连接触点分别被标定为数字 1、2 或其他数字时，可以根据 1、2 来识别路径。

normally open contact 常开触点

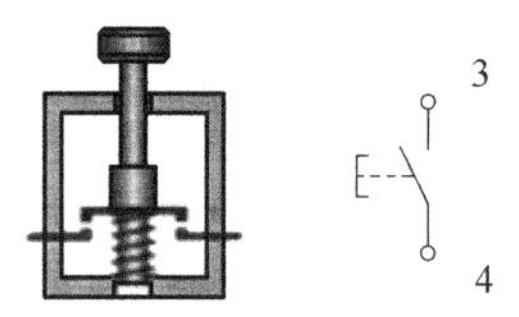

normally closed contact 常闭触点

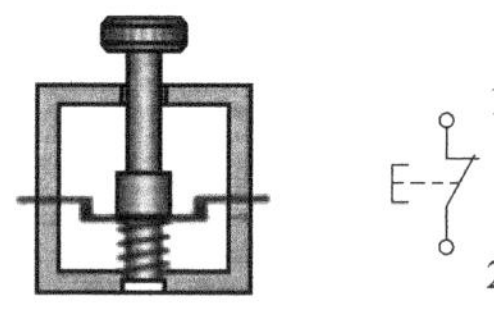

change over contact 转换触点

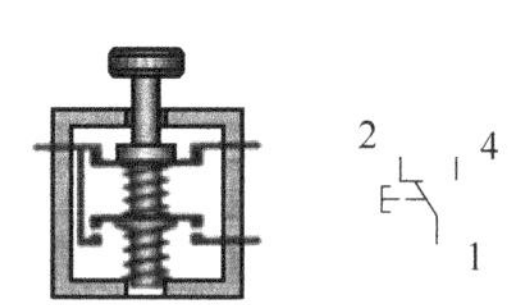

detent switch—stays on until switched in opposite direction
止动开关——保持闭合的状态直到转换到相反的方向

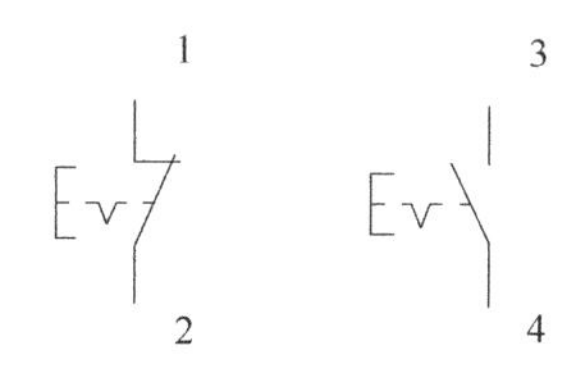

Fig. 2-14 Switch positions
图 2-14 开关位置

2. 4. 2. 3 Switch Mechanism

Limit Switch Mechanical Switches can be in a variety of shapes and sizes and depend on physical activation of the switch by contact alone, as shown in Fig. 2-15.

2. 4. 2. 3 开关机理

限位开关 机械开关具有各种形状和尺寸，并且仅通过外界的物理接触来实现其断开或闭合的功能，如图 2-15 所示。

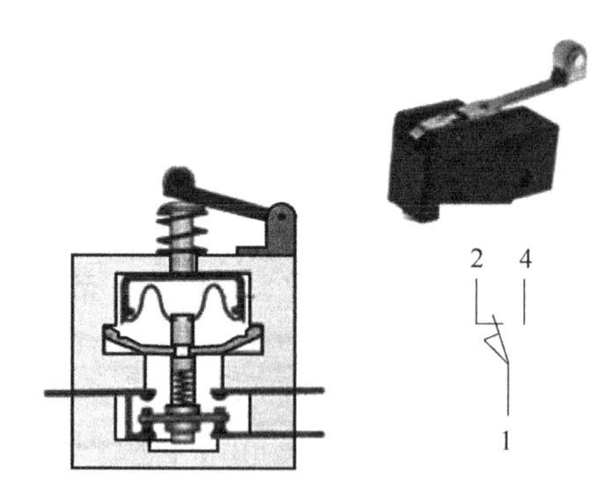

Fig. 2-15 Limit switch
图 2-15 限位开关

Reed Switch A basic sensor used in electropneumatic systems is the reed switch, as shown in Fig. 2-16. This is a simple device which is used to sense the position of the piston within the cylinder. The reed refers to a pair of contacts that are actuated by a magnetic field. The magnetic field is carried by a ring on the piston within the cylinder. Most industrial reed switches

簧片开关 簧片开关是电气系统中常用的传感器，如图 2-16 所示。它是一种用于检测活塞在气缸内所处的位置的简易装置。由簧片组成的一对触点在磁场的作用下实现其功能。磁场由位于气缸内的活塞上的环状物产生。磁敏电阻

are based on small magneto sensitive resistors which act like a switch under the influence of a magnetic field. The sensors are fitted with an indicator usually a LED (light emitting diode).

在磁场的作用下能起到类似开关的作用，因此大多数工业簧片开关都是基于磁敏电阻来实现其功能的。这种簧片开关传感器一般都配有LED（发光二极管）指示灯。

Mounting the reed switch on the cylinder is accomplished by three methods: a band around the cylinder, a rail mounting system or by attachment to a tie rod on the outside of the cylinder. Each is used according to the design of the cylinder.

将簧片开关安装到气缸上有三种方法：一种是围绕气缸形成圈状；一种是通过轨道安装系统；还有一种是在气缸外侧的拉杆上装附件。需要根据气缸的设计来决定使用哪种方法。

A reed switch can come with two or three connections. The indicator on the switch will absorb some electric current and so is usually fitted with three terminals as two connection devices are limited in the amount of current they can carry. The three terminals are as follows:

- Connection to a positive voltage rail, red or brown wire.
- Connection to a negative or zero voltage rail, blue or black wire.
- A signal output connection, black or white wire.

簧片开关可以有两个或三个接线端。开关上的指示灯需要电流才能起作用，但含有两个接线端的设备所能通过的电流量有限，因此开关上通常配备三个端子。三个终端的连接如下：

- 连接到正电压通道，红色或棕色线。
- 连接到负电压或零电压通道，蓝色或黑色线。
- 信号输出连接，黑色或白色线。

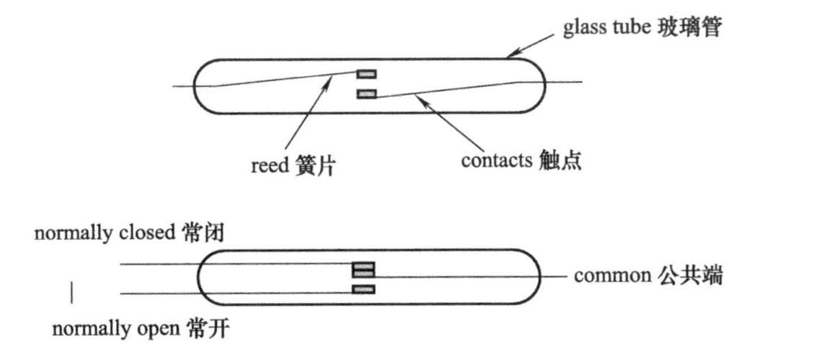

Fig. 2-16 Reed switch and symbol

图 2-16 簧片开关及其符号

2.4.3 Sensor

2.4.3 传感器

2.4.3.1 Sensor Type

2.4.3.1 传感器类型

There are a number of sensors used in electropneumatic systems with proximity sensors being the most common

电气系统中使用了许多类型的传感器，其中接近传感器是最常见的

type. The operating principles are based on a number of physical effects. Most require some electrical energy to work.

类型，其工作原理基于许多物理效应，多数需要一些电能才能工作。

Inductive Detect the presence of a metal. The switch is activated when a metal part is placed in the target zone. It must be provided with electrical power. The inductive sensor is capable of working in dirty environments even submerged in fluids. The key dimension of an inductive sensor is its diameter, e. g. , a 4mm diameter sensor has a range of 1mm but a 40mm diameter has a range of 20mm, although they are used over shorter ranges. The target material will also affect the response of the sensor. The manufacturer will specify the range and then a % range for different types of material, e. g. , mild steel 100%, Cast Iron 110%, Aluminium 35%, etc.

电感式 用于检测金属的存在。在供电的情况下，当金属部件放置在目标区域中时，开关被激活。电感式传感器能够在恶劣的环境甚至浸没在流体中工作。电感式传感器的关键尺寸是其直径，例如直径为4mm的传感器的感应距离为1mm，直径为40mm的传感器的感应距离为20mm，它们一般应用于感应距离较小的场合。目标材料也会影响传感器的响应速度，制造商通常会标明感应距离以及不同类型材料的感应距离与此感应距离的百分比，例如低碳钢 100%、铸铁 110%、铝 35%等。

Capacitive Detect the presence of most materials although the material can have a strong effect on the range of detection, e. g. , metal 65mm, PVC 20mm.

电容式 可以检测大多数材料的存在，尽管材料的属性对检测范围有很大的影响。例如，当检测目标为金属材质时，感应距离为65mm；当检测目标为PVC材质时，感应距离为20mm。

Ultrasonic Transmit a signal and catch the reflected sound. Can have ranges up to 500mm, by using separate transmitters and receivers.

超声波式 传输信号并捕捉回声。通过使用单独的发射器和接收器，其检测距离可达500mm。

Light Infrared and fibre optic sensors used in much the same way as ultrasonic sensors.

光式 红外式及光纤式传感器的使用方式与超声波式传感器大致相同。

2. 4. 3. 2 Wiring of Sensors

2. 4. 3. 2 传感器接线

Fig. 2-17 shows the symbols are used for capacitive and inductive sensors.

电容式和电感式传感器的符号如图2-17所示。

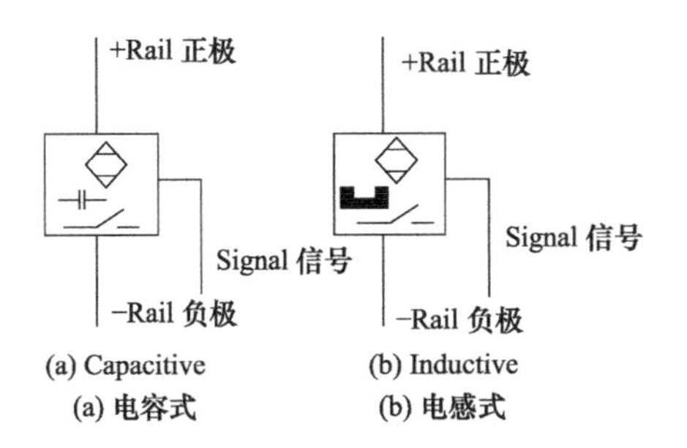

Fig. 2-17 Sensor symbols
图 2-17 传感器符号

It is important to know if a switch is a PNP or NPN type (Fig. 2-18). PNP means that the signal wire is switched to the positive rail supply, i. e., from 0V to 24V. NPN means that the signal wire is switched to the negative supply, i. e., from 24V to 0V.

了解开关是 PNP 型还是 NPN 型是非常重要的（图 2-18），PNP 型意味着信号线被切换到供应电源正极，即从 0V 到 24V；NPN 型意味着信号线被切换到供应电源负极，即从 24V 到 0V。

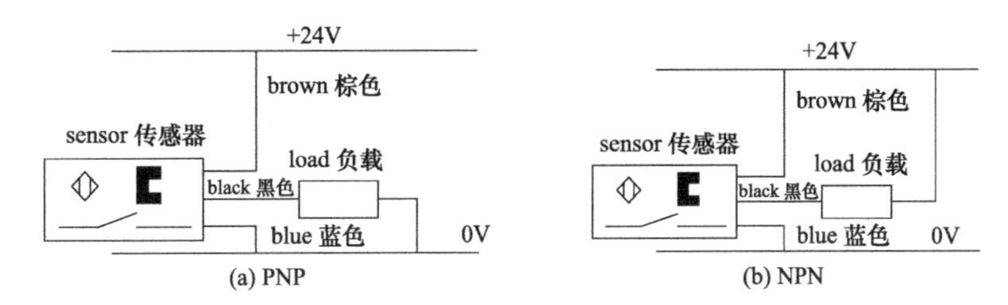

Fig. 2-18 PNP and NPN sensors
图 2-18 PNP 型和 NPN 型传感器

Connecting up the switches incorrectly can cause "burn out" of the switch. This is not always obvious as the indication light may still be illuminated.

错误地连接开关可能导致其"烧坏"，然而这种情况并不容易被发现，因为即使在这种情况下指示灯仍然可以亮。

Usually the proximity switches are limited in the amount of current that can be passed through the internal contacts. Care must be taken when connecting up the sensors directly to solenoids and possibly should be avoided.

通常，开关的使用范围受限于可以通过其内部触点的电流量。将传感器直接连接到螺线管时必须倍加小心，且应尽可能避免这种情况。

2.4.4 Relay

2.4.4 继电器

A relay is an electro mechanical device that uses a solenoid to activate the various contacts that are built into the relay.

继电器是一种采用螺线管来激活其内部各种不同触点的电动机械装置。

Fig. 2-19 shows the structure of a relay. The basic principle is that a voltage applied to the coil pulls the lever towards the coil. Voltage applied at contact 1 will be transferred to contact 4. Removal of the voltage on the coil allows the lever to return to its original position due to the spring. Electric current can then flow from contact 1 to 2. Relays are useful because they can be designed to operate multiple sets of contacts (switches poles). By having a number of contacts operated by a relay it is possible to start or end a number of operations either together or in a sequence as determined by the designer of the machine. It is also possible to switch large voltage sources to motors and other devices, allowing a small voltage to control large machines.

图 2-19 为一继电器的结构图，其基本原理是：在带电线圈的作用下，杠杆将被拉向线圈，施加于触点 1 上的电压将转移到触点 4 上；若线圈失电，杠杆将会在弹簧的作用下返回到其原始位置，那么电流将从触点 1 流向触点 2。继电器用途广泛，通过进一步设计可以操纵多组触点（开关极）。通过能够操纵多组触点的继电器，可以同时或按照设计者指定的顺序启动或结束机器的多个运作模式。它也可以将大电压电源切换到马达或其他设备上，即允许用小电压来控制大型机器。

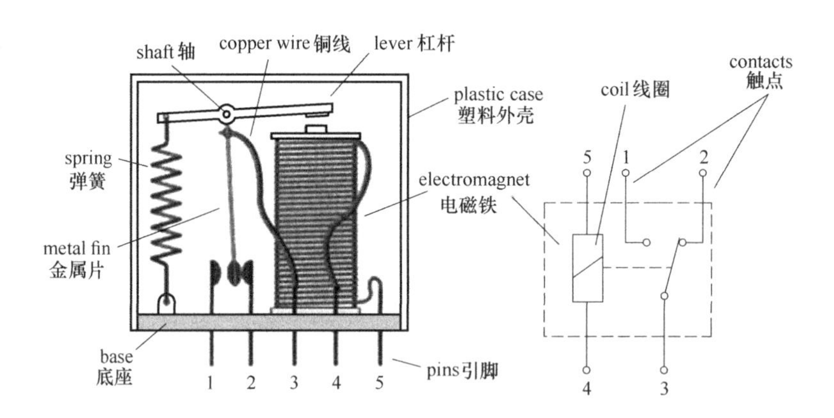

Fig. 2-19 Relay structure and symbol

图 2-19 继电器的结构及其符号

2.5 Design of Electropneumatic Circuit

2.5 电气电路的设计

An electropneumatic system uses solenoid controlled valves and an electrical control circuit. Symbols are used to represent each element of the electropneumatic system. The standard used is ISO 1219.

电气系统通常采用电磁控制阀及电气控制回路。通常用基于 ISO 1219 标准的符号来表示电气系统中的每个元件。

A symbol does not represent the following characteristics:

- Size or dimensions of the component.
- Particular manufacturer, methods of construction or costs.
- Operation of the ports.
- Any physical details of the elements.
- Any unions or connections other than junctions.

符号不代表以下特征：

- 部件的大小或尺寸。
- 特定的生产商，构造方法或成本。
- 端口的运行方式。
- 元件的任何物理细节。
- 除接头以外的任何结合或连接。

Fig. 2-20 shows an electropneumatic circuit diagram. The pneumatic diagram is at the top and the electrical diagram underneath. The guidelines for drawing an electropneumatic circuit diagram are as follows:

- Place components of similar functions in appropriate groups.
- Minimise crossing lines.
- Pneumatic energy flow from bottom to top.
- Electrical energy flow from top to bottom.

图 2-20 为电气电路图，其中气动图在顶部，电气图在气动图下面。绘制电气电路图的指导方针如下：

- 将相似功能的元件放为恰当的一组。
- 最少地产生交叉线。
- 气动能从底端流向顶端。
- 电能从顶端流向底端。

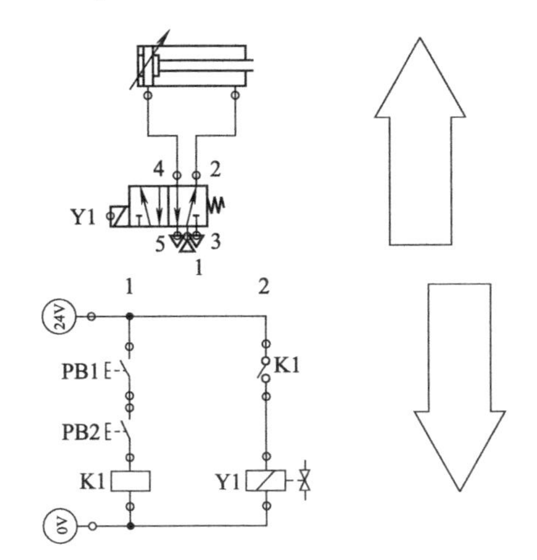

Fig. 2-20 Electropneumatic circuit diagrams

图 2-20 电气电路图

【Example 2-1】 Direct control of a single acting cylinder (Fig. 2-21).

In this case the single acting cylinder 1. 0 is controlled by a 3/2 way single solenoid valve 1. 1. A push button PB1 is used to operate the cylinder. When the push button is pressed solenoid Y1 is activated. Note that Y1 appears on both the pneumatic and the electrical diagram. The solenoid will switch the valve and air will flow from port 1 to port 2. This moves the piston and causes cylinder 1. 0 to extend, i. e. , 1. 0+ state.

【例 2-1】 直接控制单作用气缸（图 2-21）。

在这个例子中，单作用气缸 1.0 由一个 3/2 电磁阀 1.1 来控制。按钮 PB1 用于操纵气缸。当按下按钮时，螺线管 Y1 被激活。值得注意的是 Y1 在气动图和电气图上均出现。电磁阀在螺线管的作用下切换工作位置，空气将从端口 1 流向端口 2，从而使气缸 1.0 中的活塞伸出，即 1.0+状态。

When the push button is released the solenoid switches off and the spring returns the valve to its original position. The air then flows from port 2 to port 3 and is exhausted into the atmosphere. This causes the spring in cylinder 1.0 to move the piston back to its original position, i. e. , 1. 0— state.

当释放按钮时，螺线管失电，电磁阀的阀芯在弹簧的作用下回到其原始位置。然后空气从端口 2 流向端口 3 并排入大气中。从而使气缸 1. 0 中的活塞在弹簧的作用下返回其原始位置，即 1. 0—状态。

The cylinder will not move again until the push button is pressed. This is called direct control as the solenoid is activated directly by the push button.

气缸中的活塞不会再次运动，除非按下按钮。由于螺线管直接通过按钮来激活，因此称为直接控制。

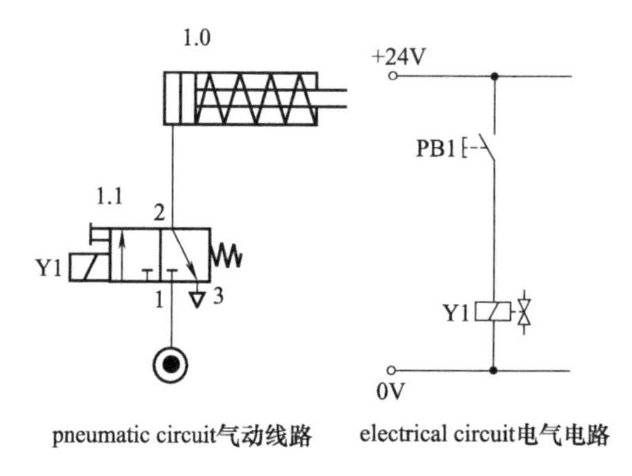

Fig. 2-21 Figure for example 2-1
图 2-21 例 2-1 附图

【Example 2-2】 Indirect control of a single acting cylinder (Fig. 2-22)

When the push button PB1 is pressed a voltage is applied to the coil of relay K1. This causes the contact K1 to close. This activates solenoid Y1 which switches the valve 1. 1 and causes the cylinder 1. 0 to extend, i. e. , 1. 0+state.

【例 2-2】 间接控制单作用气缸（图 2-22）

当按下按钮 PB1 时，电压施加到继电器 K1 的线圈上，使触点 K1 闭合。进而使螺线管 Y1 被激活，将电磁阀 1. 1 切换至工作位置，并使气缸 1. 0 伸出，即 1. 0+状态。

When the push button PB1 is released the coil of the relay is de-energised, the contact K1 opens and the solenoid goes off. Valve and cylinder return to their initial positions, i. e. , 1. 0— state. This is indirect control as the switch activates the relay and the relay contact then activates the valve.

当释放按钮 PB1 后，继电器的线圈断电，触点 K1 断开，从而使螺线管失电。电磁阀的阀芯和气缸的活塞均返回其初始位置，即 1. 0—状态。由于先通过开关激活继电器，再通过继电器的触点激活电磁阀，因此称为间接控制。

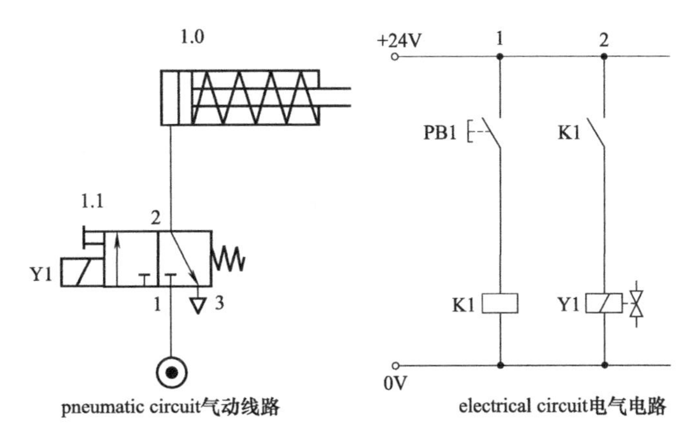

Fig. 2-22　Figure for example 2-2

图 2-22　例 2-2 附图

【Example 2-3】 Control of a double acting cylinder (Fig. 2-23).

The contacts of the pushbuttons PB1 and PB2 are arranged in parallel in the circuit diagram. When PB1 or PB2 is pressed, the coil of the relay K1 is powered up and the contact closes causing the solenoid to activate. This switches the 5/2 way valve, air flows from port 1 to port 4 and the cylinder extends (1. 0＋).

【例 2-3】 控制双作用气缸（图 2-23）

按钮 PB1 和 PB2 的触点在电路图中并联排列。当按下按钮 PB1 或 PB2 后，继电器 K1 的线圈通电，同时其触点闭合，从而使螺线管被激活。进而使 5/2 阀切换，气流从端口 1 流向端口 4，执行器伸出（1. 0＋）。

When the pushbutton is released the solenoid will become de-energised and the spring will return the valve to its initial position. Air will then flow from port 1 to port 2 to retract the cylinder and from port 4 to exhaust port 5 (1. 0－).

当释放按钮后，螺线管将断电，电磁阀的阀芯在弹簧的作用下回到其初始位置。然后空气将从端口 1 流向端口 2 从而使执行器缩回，活塞杆左侧的空气从端口 4 流向排气口 5（1. 0－）。

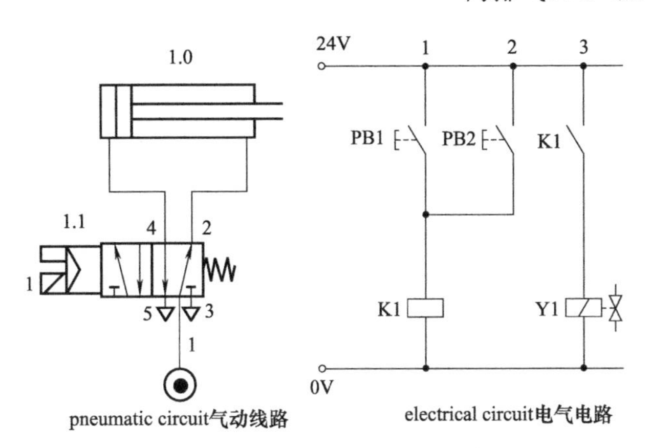

Fig 2-23　Figure for example 2-3

图 2-23　例 2-3 附图

2.6 Latch Circuit

A latch circuit is used to keep the coil of a relay powered on using one of its own contacts, as shown in Fig. 2-24.

2.6 锁存电路

锁存电路是利用继电器本身的一个触点来保持继电器接通的，如图 2-24 所示。

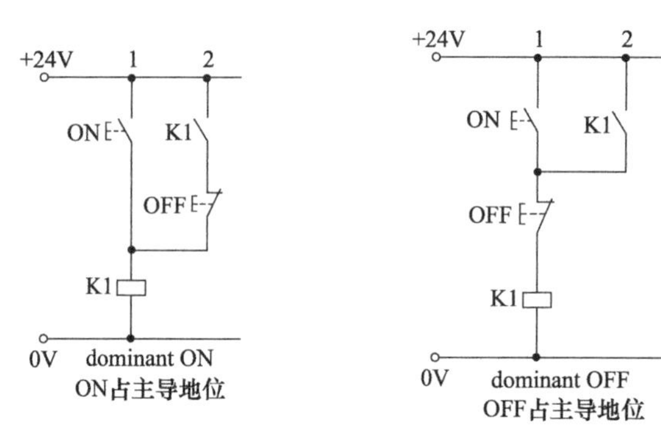

Fig. 2-24 Latch circuit

图 2-24 锁存电路

When the ON pushbutton is actuated in the circuit, the relay coil is energized and contact K1 closes. After the ON pushbutton is released, current continues to flow via contact K1 through the coil, and the relay remains in the actuated position. The ON signal is stored. This is therefore a relay circuit with a latching function. When the OFF pushbutton is pressed the flow of the current is interrupted and the relay becomes de-energised.

当电路中的 ON 按钮被按下后，继电器的线圈通电，同时其触点 K1 闭合。当 ON 按钮复位后，电流继续通过触点 K1 流过线圈，继电器保持在启动时的位置。ON 信号被存储起来。因此，这是一个具有锁存功能的继电器电路。当按下 OFF 按钮后，电流中断，同时继电器断电。

If both the ON and the OFF pushbuttons are activated at the same time and the relay coil is energized, this is a dominant ON latching circuit. If the relay coil is de-energised then it is a dominant OFF latching circuit.

如果 ON 和 OFF 按钮同时被按下且继电器线圈通电，则这是一个 ON 占主导地位的锁存电路。如果继电器线圈断电，则这是一个 OFF 占主导地位的锁存电路。

【Example 2-4】 Control of double acting cylinder—continuous action (Fig. 2-25).

When the START button PB1 is pressed, the coil K1 is powered up and all contacts on relay K1 close (rung 2, rung 3). The first contact (rung 2) on the relay closes and powers the coil K1 (rung 1).

【例 2-4】 控制双作用气缸——连续动作（图 2-25）。

当按下 START 按钮 PB1 后，继电器 K1 的线圈通电，同时继电器 K1 上的所有触点（梯级 2 和 3）闭合。继电器上的第一个触点闭合使线圈 K1（梯级 1）能够得到供电。

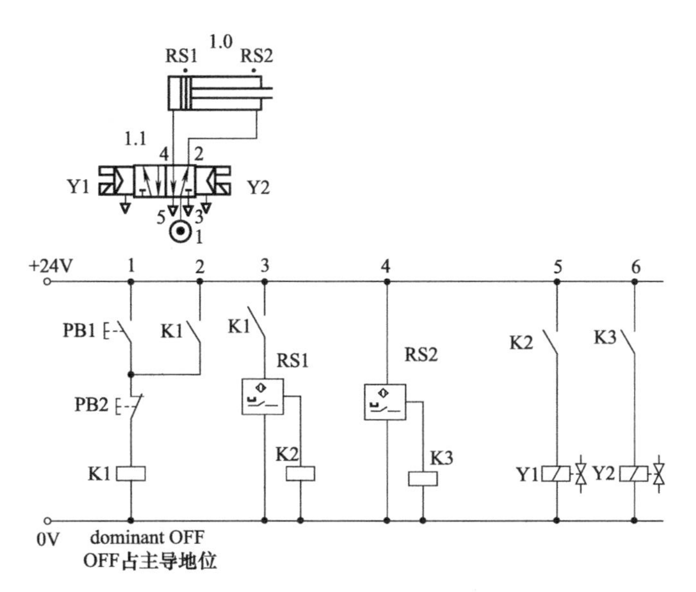

Fig. 2-25　Control of double acting cylinder—continuous action
图 2-25　控制双作用气缸——连续动作

PB1 is then released. Contact K1 (rung 2) continues to power the coil K1 (rung 1). Therefore, although PB1 is OFF the relay remains ON until the STOP button PB2 is pressed, which will break the latch. The second contact on the relay K1 (rung 3) is also closed, it supplies power to reed switch RS1 which then activates K2 (rung 3). The contact on K2 closes (rung 5) and solenoid Y1 is powered on. This switches the solenoid valve and the cylinder starts to extend (1.0+).

然后释放 PB1。继电器上的第一个触点 K1（梯级 2）仍然保持闭合状态并使线圈 K1（梯级 1）能够继续得到供电。因此，虽然 PB1 为 OFF，但继电器仍然保持为 ON，直到按下 STOP 按钮 PB2，才使锁存电路失去锁存功能。继电器 K1（梯级 3）上的第二个触点也闭合，从而使簧片开关 RS1 得到供电，进而激活 K2（梯级 3）。K2 上的触点（梯级 5）闭合使螺线管 Y1 通电。这将使电磁阀切换工作位置，执行器开始伸出（1.0+）。

As soon as the piston moves forward, the magnetic ring loses contact with the reed switch and RS1 goes off. This in turn means that K2 goes off (rung 5) and Y1 switches off. The valve remains in the same position until Y2 is powered on so the cylinder continues to extend. When the cylinder is fully extended, RS2 is activated which in turn actives Y2. This switches the solenoid valve and causes the cylinder to retract (1.0−).

当活塞一向前运动，活塞上的磁环就会与簧片开关失去感应，RS1 断开。这将使 K2（梯级 5）断开并使 Y1 关闭。电磁阀继续保持在当前的工作位置，直到 Y2 通电，因此执行器继续伸出。当执行器完全伸出时，RS2 被激活，进而激活 Y2。这将切换电磁阀的工作位置并使执行器缩回（1.0−）。

As the cylinder starts to move, it loses contact with RS2 which switches off, this switches off the relay K3 (rung 6) which switches off Y2. When the cylinder is fully retracted, RS1 will come on and if K1 (rung 2) is still latched on then Y1 will be activated and the cylinder will extend again. K1 will still be on until PB2 is pressed, breaking the latch and switching off K1 (rung 1).

当执行器开始缩回时，它将会失去与 RS2 之间的感应，从而使 RS2 关闭，进而使 K3（梯级 6）断开并使 Y2 关闭。当执行器完全缩回时，RS1 将会闭合，此时如果 K1（梯级 2）仍然保持闭合的状态那么 Y1 将会被激活，从而使执行器继续伸出。K1 将一直保持闭合的状态，直到按下 PB2，打破锁存电路并且使 K1（梯级 1）断开。

2.7 Step Diagram

2.7 阶跃图

A Step Diagram can be used to represent the movement of the cylinders during the sequence and to show when each element of the system switches on and off. This can be useful when designing a complex sequence.

阶跃图可用于反映气缸按一定顺序运动的情况，并显示系统中的每个元件何时打开和关闭，这在设计复杂序列时非常有用。

In the example shown in Fig. 2-26, the x-coordinate represents the time. The y-coordinate shows the position of cylinder 1.0 and the closing and breaking of the two relays K1 and K2, the reed switches RS1 and RS2 and the two buttons PB1 and PB2.

在图 2-26 示例中，横坐标表示时间，纵坐标显示出了气缸 1.0 的位置以及两个继电器 K1 和 K2、簧片开关 RS1、RS2 和两个按钮 PB1、PB2 闭合和断开的情况。

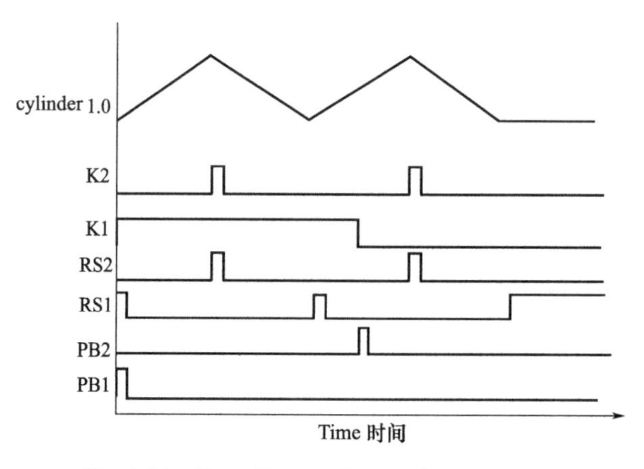

Fig. 2-26 Step diagram for 1.0+1.0− state
图 2-26 1.0+和 1.0−状态的阶跃图

Chapter 3 Programmable Logic Controller
第3章 可编程逻辑控制器

Key words 重点词汇

memory 存储器，内存
RAM 随机存取存储器
ROM 只读存储器
EPROM 可擦除可编程只读存储器
EEPROM 电可擦除可编程只读存储器
register 寄存器
shift register 移位寄存器
counter 计数器
timer 定时器
I/O imaging I/O 成像
cyclic scanning 循环扫描
phasing error 相位误差
response time 响应时间
address list 地址表
wiring diagram 接线图
ladder diagram 梯形图

At the start of industrial control most machines were controlled in some way by relays. Switches and sensors provided inputs to the machine and this information was used to control a pneumatic machine by the setting of valves in a particular sequence. This became a problem when a large number of relays and switches were used to control one machine. The relays were hard wired and the physical connection involved was very difficult to change if the machine's function needed to be changed or modified in some way.

在工业控制刚兴起时，大多数机器都是通过继电器以某种方式进行控制的。开关和传感器为机器提供输入信号，这些信息通过按特定顺序设定的阀来控制气动机器。然而，当使用大量的继电器和开关来控制一台机器时，就会带来问题。继电器的接线是固定的，如果需要以某种方式改变或修改机器的功能，则很难改变所涉及的物理连接。

This led to the development of the programmable logic controller (PLC) which replaces the physical relays in the system with a software program that reads information from the input terminals and switches on outputs according to the program and the inputs. PLC is capable of monitoring and controlling a large number of inputs and outputs (from a few dozen to a few thousand), so that even very large processes can be controlled.

这促使了可编程逻辑控制器(PLC)的发展，其用软件程序替换系统中实际的继电器，该软件程序能够从输入端读取信息并根据程序和输入来改变输出。PLC能够监视和控制大量的输入和输出（从几十到几千），因此即便是庞大的流程也能够被控制。

The control sequence performed by the PLC is determined by the program loaded into it. Each PLC can have a different program, and the program in a particular PLC can be subsequently altered, all of these makes for very flexible control.

由 PLC 执行的控制顺序是通过加载到其中的程序来决定的。每个 PLC 可以有不同的程序，特定 PLC 中的程序可以随后更改，所有这些使控制变得非常灵活。

3.1 Construction

3.1 构造

The PLC has the same basic parts as a microcomputer, i. e. , microprocessor, memory and input/output facilities (Fig. 3-1). The following is the main points that differentiate the PLC from the personal computer:

- A dedicated language is used which consists of simple logical instructions. This language is easy to learn and does not require highly trained computer personnel.
- The operating system, memory, I/O hardware and instructions are designed to meet the special requirements of logic controllers, i. e. , large number of inputs and outputs, a selection of timers, counters and flags, etc.
- Off the shelf interfaces for connecting the PLC to 110V AC or 220V AC supplies or 5V DC, 12V DC and 24V DC supplies.

PLC 具有与微型计算机相同的基本部件，即微处理器、存储器和输入/输出设备（图 3-1）。以下是区分 PLC 与个人计算机的要点：

- 使用由简单逻辑指令组成的专用语言。这种语言易于学习并且不需要训练有素的计算机人员。
- 操作系统、存储器、I/O 硬件和指令的设计旨在满足逻辑控制器的特殊要求，即大量的输入和输出、定时器的选择、计数器和标示等。
- 现成的接口，用于连接 PLC 到 110V AC 或 220V AC 电源或 5V DC、12V DC 和 24V DC 电源。

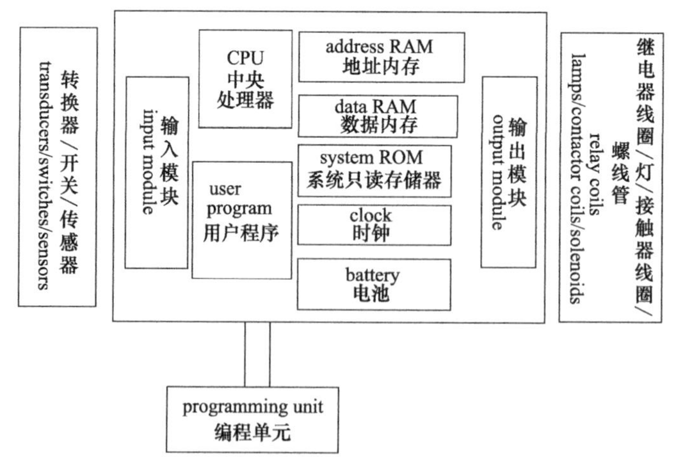

Fig. 3-1 Structure of a PLC

图 3-1 PLC 的结构

The PLC is based around a single microprocessor. The CPU is the central processing unit and this provides the means of performing the arithmetic functions. This performs the logic AND, NOT functions, etc.

PLC 基于单个微处理器。CPU 是中央处理单元，它为执行算术运算功能提供了途径，比如能够执行 AND、NOT 运算功能等。

The physical make-up of the PLC can come in two forms, rack and block types. The rack type consists of the basic unit onto which extra units can be added as required by the user. These units can take many forms, for example input units or output units, which can cater for different forms of input and output voltages, power supplies and communication units. The block type comes as a complete package, i. e. , power supply, processor, and a fixed number of inputs and outputs. This will have a limited capacity compared with the rack type.

PLC 的物理组成有两种形式，即机架型和模块型。机架型由基本单元组成，并且可以根据用户的需求添加额外的单元。这些单元可以采用多种形式，例如输入单元或输出单元，可以满足不同形式的输入和输出电压、电源和通信单元。模块型具有完整的套件，包括电源、处理器和固定数量的输入和输出，与机架型相比，容量有限。

There are many manufacturers of PLC, Omron, Siemens, Hitachi and GE are just some of the more well-known brands. Each will offer different types with a range of input and output modules for each model of PLC. Since there are no standards governing the design of PLC, this has given rise to various differences between the manufacturers concerning the programming of the PLC. These are easily overcome by the use of general programming methods.

PLC 的制造商有许多，其中欧姆龙、西门子、日立和通用电气都是一些比较知名的品牌。这些制造商都能够给每种型号的 PLC 提供不同类型的输入和输出模块。由于没有标准来主导 PLC 的设计，因此制造商之间产生了关于 PLC 编程的各种差异。通过使用通用的编程方法，可以很容易地克服这些问题。

3. 2 Memory

3. 2 内存

A memory is termed volatile if it loses data when the power to it is switched off and nonvolatile otherwise.

如果存储器在断电时丢失数据而在其他情况下不丢失数据，则称其为易失性存储器。

3. 2. 1 Memory Type

3. 2. 1 内存类型

Random access memory (RAM) Flexible type of read/write memory. Used to store ladder diagram and program data which need to be modified. RAM is volatile and cannot be used to store data when the PLC is switched off unless it is battery backed.

随机存取存储器（RAM） 读/写灵活类型的存储器，用于存储需要修改的梯形图和程序数据。RAM 是易失性的存储器，在 PLC 关闭后不能用于存储数据，除非它具有备用电池。

Read only memory (ROM) This is programmed during its manufacture. It is non-volatile and provides permanent storage for the operating system and fixed data.

只读存储器（ROM） 它在制造的过程中就被编程。它是非易失性的存储器，为操作系统和固定数据提供永久存储的功能。

Erasable Programmable Read Only Memory (EPROM) This can be programmed by electrical pulses and erased by exposure to UV light. It is a non-volatile memory, which provides permanent storage for ladder programs.

可擦除可编程只读存储器（EPROM） 可以通过电脉冲进行编程，并通过暴露于紫外线下进行擦除。它是一个非易失性存储器，为梯形图程序提供永久存储的功能。

Electrically Erasable Programmable Read Only Memory (EEPROM) This is similar to EPROM but is erased by using electrical pulses rather than UV light.

电可擦除可编程只读存储器（EEPROM） 其类似于EPROM，但是通过使用电脉冲而不是紫外线来擦除编程。

3.2.2 Memory Map

3.2.2 内存映射

Fig. 3-2 shows the allocation of memory addresses to ROM, RAM and I/O. Generally the memory map is configured by the manufacturer which means that the program capacity, the number of I/O ports and the number of internal flags, counters and timers are fixed.

图3-2所示为ROM、RAM和I/O的内存分配地址。通常，存储器映射由制造商配置，这意味着程序容量，I/O端口的数量以及内部标志、计数器和定时器的数量是固定的。

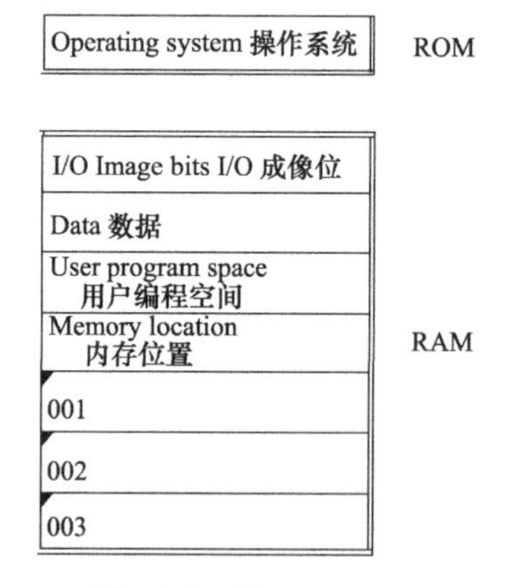

Fig. 3-2 Memory map
图3-2 存储器映射

3.3 Memory Element

3.3 存储元件

PLC memories can be thought of as large two-dimensional arrays of single unit storage cells, each storing a single piece of information in the form of 1 or 0 (i. e., the binary number format), as shown in Fig. 3-3. Each cell can only store one binary digit and is called a bit. A bit is the smallest structural unit of memory.

PLC存储器可以被认为是单个单位存储单元的大型二维阵列，每个存储单元以1或0的形式（即二进制数格式）存储单条信息，如图3-3所示。每个单位只能存储一个二进制数字，称为位。位是内存

Although each bit stores information as 1 or 0, the memory cells do not actually contain the numbers 1 and 0. The cells use voltage changes to represent 1 and 0. A bit is considered to be ON if the stored information is 1 (voltage present) and OFF if the stored information is 0 (voltage absent).

的最小结构单位。虽然每个位将信息存储为 1 或 0，但存储器单元实际上不包含数字 1 和 0，它使用电压变化来表示 1 和 0。如果存储的信息为 1（电压存在），则该位被认为是处于 ON 状态；如果存储的信息为 0（电压不存在），则被认为是处于 OFF 状态。

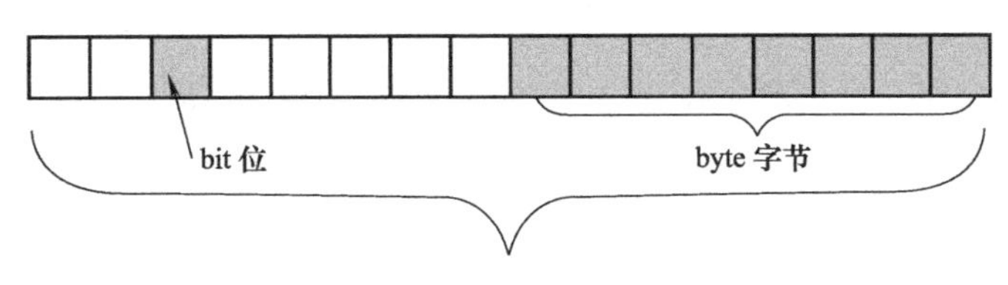

Fig. 3-3 Memory elements

图 3-3 存储单元

Sometimes a processor can handle more than a single bit of data at a time. For example, it is more efficient for a processor to work with a group of bits when transferring data to and from memory. Also storing numbers and codes requires groups of bits. The smallest group of bits that can be handled by the processor at one time is called a byte. Byte size is normally eight bits.

有时，处理器一次可以处理多个数据。例如，在向存储器传输数据和从存储器输出数据时，处理器同时处理一组位效率更高。同时存储数字和代码也往往需要很多位。处理器一次可以处理的最小位组称为字节，字节大小通常为 8 位。

Auxiliary relays They are single bit memory elements located in RAM that may be manipulated by the user program. They are called auxiliary relays because they are likened to imaginary internal relays. A battery backed auxiliary relay is called a retentive or holding relay and can be used for storing data during power failure.

辅助继电器 是位于 RAM 中的单位存储元件，可由用户操纵。它们之所以被称为辅助继电器，是因为它们被比作假想的内部继电器。电池供电的辅助继电器称为有记忆的继电器或保持继电器，可在电源故障期间存储数据。

Register Another information unit used by the PLC is a register. A register is the unit that the processor uses when data is to be operated on or instructions are to be performed. Typical register sizes are 8, 16 and 32 bits. Most CPU operations involve the use of a register. The types of register include:

寄存器 PLC 使用的另一个信息单元是寄存器。通常，寄存器是处理器在运算数据或执行指令时使用的单元。典型的寄存器的大小有 8、16 和 32 位。大多数 CPU 的运算都需要使用寄存器。寄存器的类型有：

① Data registers: They are located in RAM and are used for storing flags, counters and timers constants and other types of data.

① 数据寄存器：位于RAM中，用于存储标志、计数器和定时器常数以及其他类型的数据。

② Flag registers: If a bit state is used to indicate that a condition has occurred it is called a flag. The CPU has an internal flag register which contains information about the result of the latest arithmetic and logical operations.

② 标志寄存器：如果使用位状态来标识已经发生的条件，则将其称为标志。CPU有一个内部标志寄存器，其中包含关于最后一次算法和逻辑运算结果的信息。

③ Shift Registers: Some registers are arranged so that bits stored in them can be moved one position to the left or right, these are called shift registers and can be used for sequence control applications.

③ 移位寄存器：一些寄存器用来使存储在其中的位向左或向右移动，这种寄存器称为移位寄存器并可以用于顺序控制的场合。

Auxiliary components The normal operation of the register is inseparable from the following components.

辅助元件 寄存器的正常工作离不开以下元件的配合。

① Binary counters: The CPU is able to function as a binary counter since it is able to increment or decrement binary data stored in a register and compare binary data stored in two separate registers. Counters are used to count, for example, digital pulses generated from a switching device connected to an input port. The count value required is stored in a data register.

① 二进制计数器：CPU能够用作二进制计数器，因为它能够使存储在寄存器中的二进制数据递增或递减并能够比较存储在两个独立寄存器中的二进制数据。计数器用于计数，例如计算连接到输入端口的开关设备产生的数字脉冲数。所需的计数值存储在数据寄存器中。

② Timers: A CPU will have a built-in clock oscillator which controls the rate at which it operates. The CPU uses the clock signal to generate a time delay.

② 定时器：CPU有一个内置的时钟振荡器，用于控制其工作速率。CPU使用时钟信号来产生一个时间延迟。

3.4 Operation of the PLC

3.4 PLC的运行

The information in the address and data RAM is defined by the user who has entered a program via the programming unit in a series of steps. These steps define the order of the program. The address refers to a

地址以及数据RAM中的信息由用户来定义，用户可以按照一系列步骤通过编程单元进入程序。这些步骤定义了程序的运行顺序。地

particular input number or output number and the data is the state of the inputs, outputs, counters, timers as set by the user or read in from the inputs. As the CPU works through each line of the program the various outputs, timers or counters will be set according to the program. The operation of the PLC is controlled by a clock, which acts on the CPU. Each pulse of the clock causes the CPU to cycle through one step of the program once the program is set to run.

址是指特定的输入编号或输出编号，数据是由用户设置或从输入端读取的输入、输出、计数器、定时器的状态。当 CPU 每运行一行程序时，将根据程序设置各种输出、定时器或计数器。PLC 的运行由时钟控制，该时钟也对 CPU 起作用。一旦程序运行，每经过一个时钟的脉冲，都会使 CPU 循环执行程序的一个步骤。

3.4.1 PLC Operating System

All PLC operating systems execute a ladder diagram by scanning one rung at a time sequentially. The inputs of the first rung are scanned and the logic solved to determine the logic state of its output, i. e. , whether the output should be ON or OFF. This process is repeated for the second and third rung. When the "end of program" line is reached the scan cycle repeats itself so that each rung is scanned over and over again. If there is no "end of program" line, the PLC must scan through all possible steps before beginning a new scan. The program logic may involve simple AND, OR and NOT functions or more advanced timing, counting, sequence and mathematical functions. The operating system determines what functions are available to the user.

3.4.1 PLC 操作系统

所有的 PLC 操作系统都是通过按时间顺序扫描每一个梯级来完成一个梯形图的。扫描第一行梯级并得到输入，解析其相应的逻辑，以确定其输出的逻辑状态，即输出应该是 ON 还是 OFF。对第二和第三梯级重复该过程。当达到“程序结束”行时，重复运行整个扫描周期以便反复扫描每个梯级。如果没有“程序结束”行，PLC 必须在重新开始扫描之前扫描所有可能的步骤。程序逻辑可以涉及简单的 AND、OR、NOT 函数或更高级的定时、计数、序列和数学函数。操作系统可以决定用户能够使用哪些功能。

3.4.2 Program Execution

Fig. 3-4 shows the sequence used by the PLC when running a program. The CPU carries out internal processing which does things such as checking that the program structure is correct and reading values of timers and counters. Then the inputs are read. If the PLC is in run mode, the PLC will work through each line of code and decide if the output should be ON or OFF.

3.4.2 程序执行

图 3-4 显示了 PLC 在运行一个程序时的顺序。首先，CPU 执行其内部进程，例如检查程序结构是否正确以及读取定时器和计数器的值，然后读取输入信息。如果 PLC 处于运行模式，它将会运行每一行代码并确定输出是 ON 还是

When each line of code has been scanned, the PLC will then update the outputs. The process will begin again. If the PLC is in stop mode, program execution does not occur.

OFF。当PLC扫描每行代码时，将会同时更新其输出信息。以上过程将会重复进行。如果PLC处于停止模式，则不会再次进行该过程。

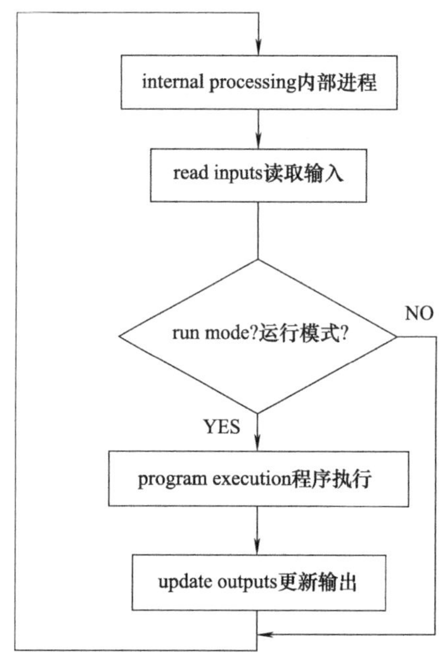

Fig. 3-4 Program execution

图 3-4 程序的执行

Thousands of different inputs and outputs can be connected to a PLC used to control a large automated system. The CPU, however, can only cope with one instruction at a time. During program execution, therefore, the status of the different input signals must be read off one at a time. The result of the different logical operations on the input signals has to be written out to the various output channels. To enable rapid program execution, input and output feeding is carried out at regular time intervals. This process, also known as I/O imaging, is done automatically by the CPU.

数千种不同的输入和输出可以连接到一个用于控制大型自动化系统的PLC上。但是，CPU一次只能处理一条指令。因此，在程序执行期间，每次执行程序时，不同输入信号的状态必须被读取一次。同时，输入信号的不同逻辑运算结果必须分别输出到不同的输出通道。为了能够快速执行程序，输入和输出供给是按照固定的时间间隔完成的，此过程也称为I/O成像，由CPU自动完成。

3.4.3 I/O Imaging

3.4.3 I/O成像

Fig. 3-5 shows the principle of I/O imaging. A special memory (I/O RAM) is used as a buffer between the

图3-5展示了I/O成像的原理。一个特殊存储器（I/O RAM）被用作

control program and the I/O unit. A memory cell (bit) in the I/O RAM represents each input and output.

控制程序和 I/O 装置之间的缓冲区。I/O RAM 中的存储单元（位）表示每个输入和输出。

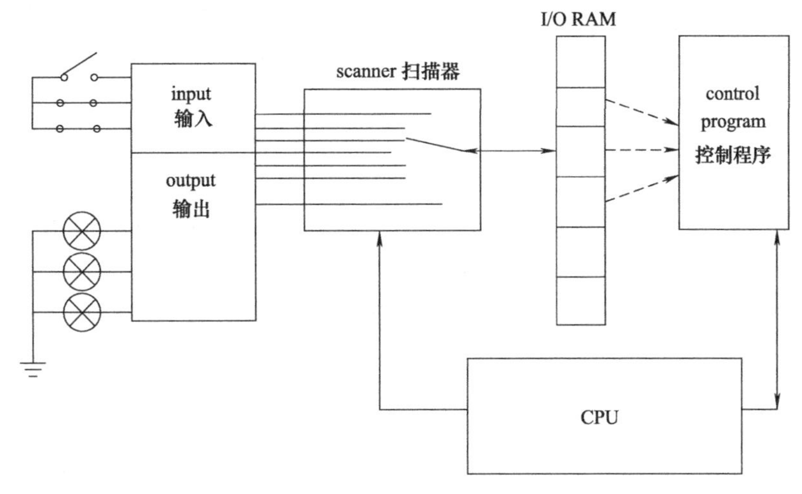

Fig. 3-5 Principle of I/O imaging

图 3-5 I/O 成像原理

The process is as follows:

其过程如下：

- The CPU scans all the inputs.

- CPU 扫描所有的输入。

- It updates the input RAM by putting a logic 1 in the bit representing an input if that input is ON and a logic 0 if the input is OFF. For example, if the sensor connected to input I0. 1 is ON then the bit for I0. 1 will have a logic 1.

- 如果输入为 ON，则通过把这个输入的位置为逻辑 1 来更新输入 RAM；如果输入为 OFF，则将这个输入的位置为逻辑 0。例如，如果连接到输入 I0. 1 的传感器为 ON，那么表示 I0. 1 的位将置为逻辑 1。

- The first rung of the ladder is then scanned using the recorded input data and the condition of timers, counters and outputs and the decision made to turn on or off the output.

- 然后利用所记录的输入数据，定时器、计数器、输出的情况和打开或关闭输出的决定，对梯形图的第一个梯级进行扫描。

- If the output should be turned ON, a logic 1 is placed in the corresponding bit in the I/O RAM. For example, if output Q0. 1 should be on then the bit representing Q0. 1 will have a logic 1.

- 如果输出应为 ON，则将 1 放置在 I/O RAM 的相应位中。例如，如果输出 Q0. 1 应为 ON，则代表 Q0. 1 的位将置为逻辑 1。

• Each rung of the ladder is scanned in turn and the I/O RAM updated.

• The status of the outputs are then updated to reflect the I/O RAM.

• Step 1 is repeated.

•依次扫描梯形图的每个梯级并更新 I/O RAM。

•然后更新输出的状态以反映 I/O RAM 的变化。

•重复步骤 1。

3.4.4 Cyclic Scanning

3.4.4 循环扫描

Cyclic scanning begins with internal processing, followed by I/O imaging, and then the different instruction sequences are executed in the order they occur in the program. A new I/O imaging follows the final line, after which the program is again executed and so on. The total cycle time, i.e., the time for the program to complete a memory cycle, depends on the size of the control program. A program of 1kB, depending on the type of PLC system, has a cycle time of between 1 and 10 milliseconds.

循环扫描从内部进程开始，接着是 I/O 成像，然后按照它们在程序中出现的顺序执行不同的指令序列。最后一行后面有一个新的 I/O 成像，之后再次执行这个程序，依此类推。总循环时间，即程序完成一个内存循环的时间，取决于控制程序的规模。根据 PLC 系统的类型，1kB 的程序的循环时间一般为 1～10 毫秒不等。

An important consequence of the execution of a cyclic program is that the status of an input signal in the I/O RAM cannot be altered within one and the same program cycle. In the same way, the status of an output signal in the I/O RAM can be altered within the program cycle without changing the value of the actual output. Only when a whole program cycle is completed the I/O RAM feed the I/O unit with the latest values. These two points are important to remember when programming a PLC system.

循环程序执行的重要推论是不能在同一个程序循环周期内改变位于 I/O RAM 中的某个输入信号的状态。同样的，在不改变实际输出值的情况下，可以在程序循环周期内改变 I/O RAM 中输出信号的状态。只有在整个程序循环完成后，I/O RAM 才会为 I/O 单元提供最新的值。在进行 PLC 系统编程时，这两点非常重要。

The operating system is characterized by the following aspect.

操作系统的特点由以下几个方面决定。

Scan Rate This is the speed with which the PLC scans the memory. The scan rate depends how fast the CPU is clocked. It is expressed in how many seconds it takes to scan a given amount of memory, usually 1kB. The actual time to scan a program will depend on

扫描速率 是指 PLC 扫描内存的速度。扫描速率取决于 CPU 计时速率的快慢，它以扫描给定内存（通常为 1kB）所需的秒数表示。扫描一个程序的实际时间取决

the scan rate, the length of the program and the types of functions used in the program.

于扫描速率、程序长度和程序中使用函数的类型。

Phasing Errors The CPU under the control of the operating system scans the input image memory rather than the inputs themselves. The input memory is not changed while the CPU is scanning it so it is possible for an input port to change its state before the input image memory is updated. A phasing error is said to have occurred when the CPU scan misses a change of state of an input port.

相位误差 操作系统控制下的CPU扫描的是输入镜像存储器而不是输入本身。当CPU在扫描时，输入存储器不会改变，因此，在输入镜像存储器更新之前，输入端口可以改变其状态。当CPU在扫描过程中错过某个输入端口的状态变化时，就会出现相位误差。

Response Time The response time of the PLC is the delay between an input being turned on and the output changing state. Delays can be due to:

- The mechanical response of an output device such as a relay.
- The electrical response of an input circuit.
- The scan update of image memory.

响应时间 PLC的响应时间是输入打开和输出状态改变之间的延迟时间。延迟可能是由于：

- 输出设备（如继电器）的机械响应。
- 输入电路的电气响应。
- 镜像存储器的扫描更新。

3.5 Input/Output (I/O) Module

3.5 输入/输出（I/O）模块

A PLC can have differing numbers of inputs and outputs, for example 20, 40 up to 5000 I/O and more. This can be confusing, as it is not self-evident as to the actual number of inputs and the actual number of outputs. For example, a 20 I/O PLC may have 8 outputs and 12 inputs or 10 inputs and 10 outputs. All PLC manufacturers differ somewhat in their respective configurations; therefore it is important to quantify the actual requirements of inputs and outputs. The inputs and outputs are real actuators and devices.

通常，PLC具有不同数量的输入和输出，例如20、40，甚至5000或更多个I/O。实际输入和输出的数量并非是明确的，这可能会造成混乱。例如，具有20个I/O的PLC可能有8个输出和12个输入或10个输入和10个输出。在PLC的配置方面，所有的PLC制造商在各自产品的配置上均有不同，因此，量化输入和输出的实际需求非常重要。输入和输出是实际的执行器和设备。

Table 3-1　Typical inputs and outputs

表 3-1　典型的输入与输出

inputs 输入	outputs 输出
ON/OFF switches 启动/停止开关	
push buttons 按钮	starters 启动器
overload signal contacts 超载信号接触	pilot lights 飞行灯
limit switches 限位开关	relays 继电器
float switches 浮动开关	solenoid valves 电磁阀
plow switches 流量开关	alarms 报警器
proximity sensors 接近传感器	motors 马达
photoelectric sensors 光电传感器	

Table 3-1 above gives an indication of the types of inputs and outputs used. The hazards of faults, burn outs, etc., may be happened for these inputs and outputs. It is for this reason that all I/O are protected individually by opto-couplers or fuses, thus protecting the PLC instantly against external faults.

表 3-1 给出了一些典型输入和输出类型的示例。这些输入、输出可能发生的故障、烧坏等危险。因此，所有的 I/O 都由光耦合器或熔断器单独保护，从而使 PLC 能够立即应对外部故障。

The internal signal level in a PLC usually is between 5～15 V. The signal level outside the PLC is usually higher, generally between 24～240 V. All communication to be carried out between the PLC and the process is done via the I/O unit, which converts the process signal to the PC system signal level. In this way, a PLC can often be used to drive relays and solenoid valves directly, without the need for intermediate relays.

PLC 内部信号电平通常介于 5～15V 之间；PLC 外部信号电平通常较高，一般在 24～240V 之间。PLC 和进程之间的所有通信都是通过 I/O 单元来执行的，I/O 单元将进程信号转换为 PC 系统的电平信号。通过这种方式，PLC 通常可以用于直接驱动继电器和电磁阀，无需中间继电器。

All transducers and actuators are connected directly to the I/O unit. On connection each signal is assigned to its own individual channel number, also called its address. The channel number is used during programming when the status of a transducer is required or a particular output is to be changed. Most I/O units are in modular form. An I/O module, usually consisting of a single circuit board, accepts connections from several input and output signals of the same type. Various I/O modules (circuit boards) are designed to fit into the rack, which in most PLC systems also contains the CPU and the memory.

所有传感器和执行器都直接连接到 I/O 单元。在连接时，每个信号分配有自己的通道号，这些通道号也称为地址。在编程期间，当需要知道传感器的状态或要改变某个特定的输出时，就会用到这些通道编号。大多数 I/O 单元采用模块化形式。I/O 模块通常由单个电路板组成，可接受多个具有相同类型的输入和输出信号的连接。通过设计各种 I/O 模块（电路板），使其可以安装在不同的机架中，在大多数 PLC 系统中，I/O 模块也包含 CPU 和内存。

3. 5. 1 Input Module

Each input signal passes through a status LED (for signal indication on the module), a voltage dropping resistor and an opto-isolator. In an AC module it also passes through a rectifier. The objective is to provide the CPU with a signal it can handle, 5V DC free of interference and electrically isolated. Fig. 3-6 shows a typical input module.

3. 5. 1 输入模块

每个输入信号都会通过一个状态LED（用于模块上的信号指示）、一个降压电阻和一个光隔离器。在AC模块中，它还通过整流器，目的是为CPU提供可以处理的5V DC无干扰信号以及进行电气隔离。图3-6为一典型的输入模块。

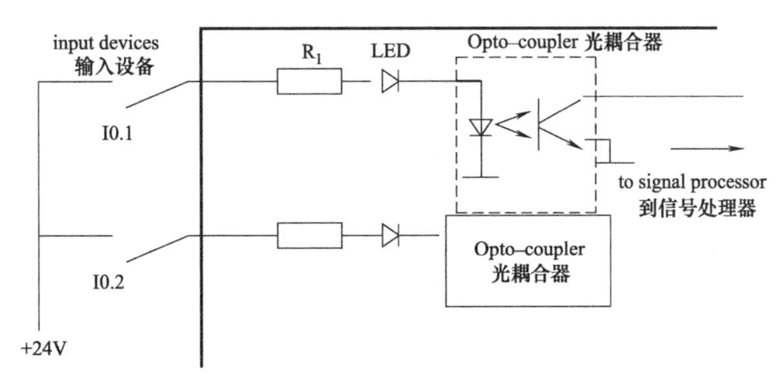

Fig. 3-6 Input module

图 3-6 输入模块

An opto-coupler (also called opto-isolator) is a device which uses light to couple signals from one system to another. In this case, the input device and the input image memory circuit. It incorporates a light emitting diode (LED) and phototransistor for this purpose. The device provides a very large degree of electrical isolation between two circuits.

光耦合器（也称为光隔离器）是一种利用光将信号从一个系统耦合到另一个系统的装置。在这种情形下，输入设备和输入镜像为存储电路，与发光二极管（LED）和光电晶体管组合在一起。该器件能够在两个电路之间提供非常好的电气隔离效果。

When a voltage is applied to the input port, current flows through the opto-coupler LED. This causes it to transmit light and turn on the phototransistor. This allows the input signal to be transmitted without any transfer of voltage.

当电压施加到输入端口时，电流流过光耦合器LED。这使其传输光并打开光电晶体管。从而在没有任何电压传输的情况下进行输入信号的传输。

3. 5. 2 Output Module

The output module of a PLC transfers the output signals from the microprocessor at a very low voltage to

3. 5. 2 输出模块

PLC的输出模块能够将来自微处理器的输出信号以非常低的电压传

the output terminals of the controller at plant voltage, e. g. , 110/220V AC. Opto-couplers provide electrical isolation and limit the effects of interference, and an LED provides an indication.

送到具有高电压的控制器的输出端，例如输出端电压为 110/220V AC 的厂用电压。光耦合器能够提供电气隔离并限制干扰的影响，其中 LED 用于指示作用。

The output module in Fig. 3-7 is a relay type output module. When the signal for the processor indicates that the output should be on, the information is passed through the opto-coupler and the indicating LED and then to the coil of the relay K1. When the relay coil is powered up, the contact closes and the 24V supply connected to the common activates the solenoid connected to Q0. 1, switching on the output.

图 3-7 为继电器型输出模块。当处理器的信号指示输出应该打开时，信息通过光耦合器和 LED 指示灯到达继电器 K1 的线圈。当继电器的线圈通电时，其触点闭合，公共端的 24V 电源将连接到 Q0. 1 的螺线管激活，从而使输出打开。

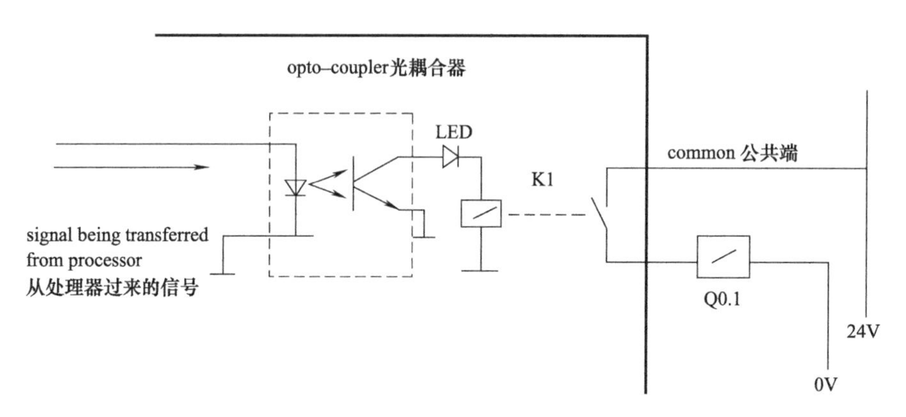

Fig. 3-7 Output module
图 3-7 输出模块

3. 6 Address List and Wiring Diagram

3. 6 地址表和接线图

When designing and building an automated system the first step is to decide what inputs and outputs are needed and then to decide where they will be connected to on the PLC. This is called an address list. For example, in a pneumatic system there may be reed switches (RS1, RS2 etc.), push button switches (PB1, PB2 etc.) and solenoids (Y1, Y2 etc.) connected to the PLC. The reed switches and push button switches are inputs so will be connected to the input module of

在设计和构建一个自动化系统时，首先要确定需要哪些输入和输出，然后再确定它们在 PLC 上的连接位置，表示这种连接关系的表格称为地址表。例如，在气动系统中，可能有连接到 PLC 的簧片开关（RS1，RS2 等)、按钮开关(PB1，PB2 等）和螺线管（Y1，Y2 等)。簧片开关和按钮开关是输入，

the PLC. The solenoids are outputs and will be connected to the output module of the PLC. So, the address list would be as Table 3-2.

因此它们将连接到 PLC 的输入模块；螺线管是输出，其将会连接到 PLC 的输出模块。因此，地址表如表 3-2 所示。

Table 3-2 Address list

表 3-2 地址表

inputs 输入	outputs 输出
RS1-I0.1	Y1-Q0.1
RS2-I0.2	Y2-Q0.2
PB1-I0.3	
PB2-I0.4	

Fig. 3-8 shows a wiring diagram for a PLC. This shows two reed switches, two push button switches and two solenoids, as identified in Table 3-2, connected to the PLC.

图 3-8 为 PLC 系统的接线图。其中有两个簧片开关、两个按钮开关和两个螺线管连接到 PLC 中，如表 3-2 所示。

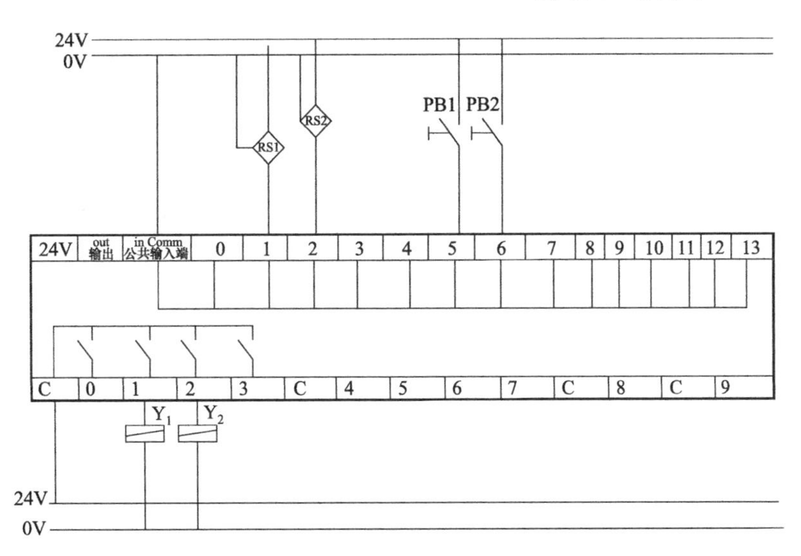

Fig. 3-8 Wiring diagram

图 3-8 接线图

Note that on the input side the common is connected to 0V as the switches are supplied with 24V. On the output side the common is connected to 24V. This supplies 24V to the relay switches in the output module. When the program activates Q0.1 the relay contact closes and 24V is supplied to the solenoid.

值得注意的是，在输入端，公共线一般连接到 0V，因为需要供给开关 24V 电压；在输出端，公共线连接到 24V，从而为输出模块中的继电器开关提供 24V 电压。当程序激活 Q0.1 时，继电器的触点闭合，从而使 24V 电压供给给螺线管。

Chapter 4 Logic System and PLC Programming
第4章 逻辑系统和PLC编程

Key words 重点词汇

logic circuit 逻辑电路	NOT gate 非门
truth tables 真值表	NAND gate 与非门
AND gate 与门	NOR gate 或非门
OR gate 或门	XOR gate 异或门

Before the advent of PLCs, a method of presenting and also understanding the behaviour of interconnected relays and switches was developed. This is used for the building of logic circuits to control simple machines and exploiting program used in PLCs.

在 PLC 出现之前，有一种用于表示和理解相互联系的继电器和开关特点的方法，以此来构建逻辑电路，以控制简单的机器和开发用于 PLC 的程序。

4.1 Logic Circuit

4.1 逻辑电路

4.1.1 AND Gate

4.1.1 与门

Consider the simple circuit shown in Fig. 4-1. It consists of two switches and a lamp.

以图 4-1 中所示的简单电路为例，它由两个开关和一个灯组成。

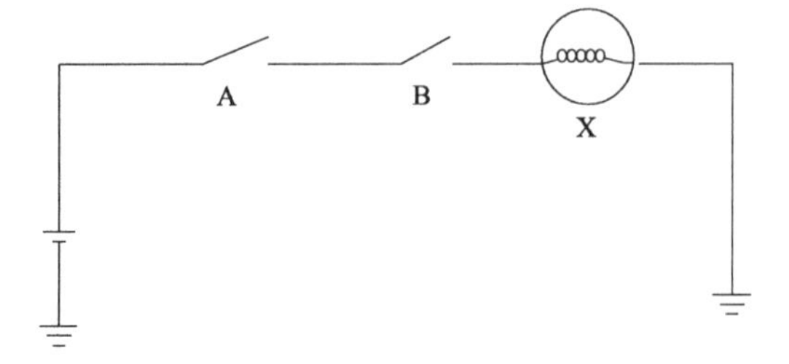

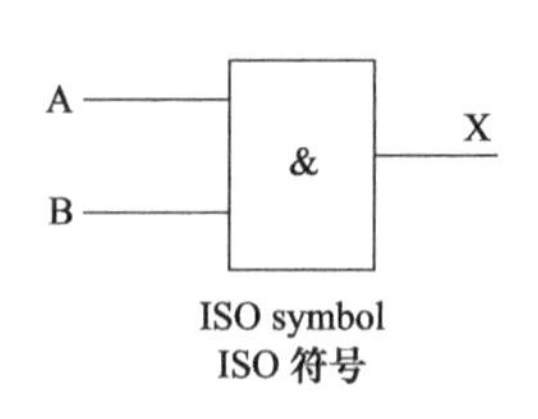

Fig. 4-1 Circuit diagram and ISO symbol for AND gate

图 4-1 与门的电路图和 ISO 符号

From the Fig. 4-1 it can be seen that lamp X will only come on if the two switches (A and B) are both closed at the same time. The switches can either be controlled by manual modes or be part of a relay switch or a

从图 4-1 中可以看出，只有当两个开关（A 和 B）同时闭合时，灯 X 才会亮起。这些开关可以通过手动方式控制，也可以是继电器开关或

proximity sensor. A very simple table can be drawn to indicate the various combinations of input and output conditions.

接近传感器的一部分。我们可以绘制一个非常简单的表来列出输入和输出条件的各种组合。

Table 4-1 Truth table for the AND gate

表 4-1 与门的真值表

input 输入		output 输出
A	B	X=A AND B X=A 与 B
0	0	0
0	1	0
1	0	0
1	1	1

Table 4-1 shows all conditions that can exist for the AND gate inputs, i. e. the positions of switches A and B, and the output of the system (lamp X on or off) for these inputs.

表 4-1 展示了与门输入可能存在的所有情况，即开关 A 和 B 的位置以及这些输入情况下的系统输出（灯 X 开或关）。

Table 4-1 is known as the truth table for the AND gate. The truth table is a method which allows the output to be divinable in terms of the inputs. The number of rows in the table is 2^n, where n is the number of inputs since each input has two possible states (i. e., 1 or 0). This means that 2 inputs requires 4 rows, 3 inputs 8 rows, etc.

表 4-1 称为与门的真值表。真值表是一种根据输入来预测输出的列表。由于每个输入有两个可能的状态（即 1 或 0），因此表中的行数是 2^n，其中 n 是输入数。这意味着 2 个输入需要 4 行，3 个输入需要 8 行等。

From the AND truth table it can be seen that, the output column (X) is the product of the two inputs (A and B), i. e., the output of the AND gate is expressed as the logical product and is represented by

从与门的真值表可以看出，输出列（X）是两个输入（A 和 B）的乘积，即与门的输出表示为逻辑乘积，表示为

$$X = A\ \text{AND}\ B = A \cdot B \tag{4-1}$$

If several inputs A, B, C, D, and so on are used then the output is the logical product of the inputs

如果有多个输入 A，B，C，D，…，则输出是输入的逻辑乘积，有

$$X = A\ \text{AND}\ B\ \text{AND}\ C\ \text{AND}\ D\ \text{AND} \cdots = A \cdot B \cdot C \cdot D \cdot \cdots \tag{4-2}$$

Eq. 4-1 and eq. 4-2 are known as Boolean expressions. The logical operator AND is denoted by a dot as shown above.

式 4-1 和式 4-2 称为布尔表达式。逻辑运算符 AND 由一个点来表示，如上所示。

4.1.2 OR Gate

Now consider a different circuit as shown in Fig. 4-2. This represents an OR gate as the output X is made by either switch A on OR switch B on.

4.1.2 或门

图 4-2 所示为一个与与门电路不同的电路。由于输出 X 由开关 A 或开关 B 控制，因此这种逻辑电路称为或门。

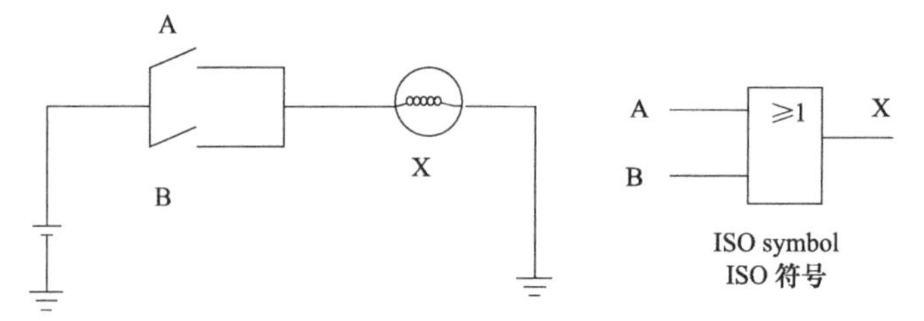

Fig. 4-2 Circuit diagram and ISO symbol for OR Gate

图 4-2 或门的电路图和 ISO 符号

Table 4-2 gives the truth table for the OR gate. From this table we can see that the output column is the algebraic sum of the two inputs columns, except for the last row of the table. Therefore, the output of the OR gate is the logical sum of the inputs. This is expressed as

表 4-2 给出了或门的真值表。从此表可以看出，除了表的最后一行，输出列是两个输入列的代数和。因此，或门的输出是输入的逻辑和，即

$$X=A \text{ OR } B=A+B \tag{4-3}$$

The OR is represented by the '+' symbol as shown above. If there are several inputs such as A, B, C, D and so on then the output can be expressed as

或由符号'+'来表示，如上所示。如果有多个输入，如 A，B，C，D，…，那么输出可以表示为

$$X=A \text{ OR } B \text{ OR } C \text{ OR } D \text{ OR}\cdots=A+B+C+D+\cdots \tag{4-4}$$

The maximum of output will still be 1. Not the total sum of the inputs but the logical sum.

输出的最大值仍然是 1，计算的不是输入的总和而是逻辑和。

Table 4-2 Truth table for the OR gate

表 4-2 或门的真值表

input 输入		output 输出
A	B	X=A OR B　X=A 或 B
0	0	0
0	1	1
1	0	1
1	1	1

4. 1. 3 NOT Gate

This Boolean function gives the inverse or negation of a variable, i. e. , if the variable A is 1 then the output is 0, if A is 0 then the output is 1. This is indicated by a bar over the variable input variable, i. e, $\overline{A}$. A circuit for such a gate is shown in Fig. 4-3.

4.1.3 非门

非门的布尔函数表示与变量相反或否定的结果，即如果变量 A 为 1 则输出为 0，如果 A 为 0 则输出为 1。通常在输入变量顶部加上一根横线来表示非门，即 $\overline{A}$。这种非门的电路如图 4-3 所示。

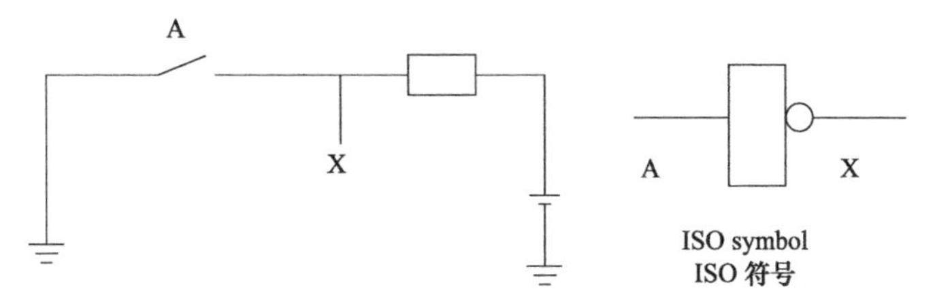

Fig. 4-3 Circuit diagram and ISO symbol for NOT gate

图 4-3 非门的电路图和 ISO 符号

If the switch A is open (i. e. , A=0), then X will be at the high voltage (i. e. , X=1). If the switch A is closed (i. e. , A=1), then X will be at the low voltage (i. e. , X=0). The truth table for the NOT gate is given in Table 4-3.

如果开关 A 断开（即 A=0），则 X 将处于高电压状态（即 X=1）。如果开关 A 闭合，(即 A=1)，则 X 将处于低电压状态（即 X=0）。表 4-3 给出了非门的真值表。

Table 4-3 Truth table for a NOT gate

表 4-3 非门的真值表

input 输入	output 输出
A	$X=\overline{A}$ X=非 A
0	1
1	0

4. 1. 4 NAND Gate

This gate is made up of a NOT gate and an AND gate. The logical output of this gate is the inverse of the AND gate. Its symbol is a combination of the two gate symbols, as shown in Fig. 4-4.

The truth table for this gate is given in Table 4-4. We can see that the output of the NAND gate is the logical inverse of the AND gate result.

4.1.4 与非门

该门由非门和与门组成，其逻辑输出与与门相反。与非门的符号是两个门符号的组合，如图 4-4 所示。

表 4-4 给出了该门的真值表。从表中可以看出，与非门的输出与与门的输出结果是逻辑相反的。

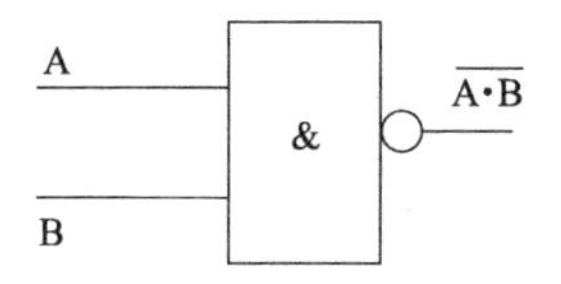

Fig. 4-4　NAND gate symbol

图 4-4　与非门符号

Table 4-4　Truth table for NAND gate

表 4-4　与非门的真值表

input 输入		output 输出	
A	B	AND 与	NAND 与非
0	0	0	1
0	1	0	1
1	0	0	1
1	1	1	0

4. 1. 5　NOR Gate

4. 1. 5　或非门

One other useful gate is the NOR gate (Not Or). The symbol for this is given in Fig. 4-5.

另一个常用的逻辑门是或非门(Not Or)，其符号如图 4-5 所示。

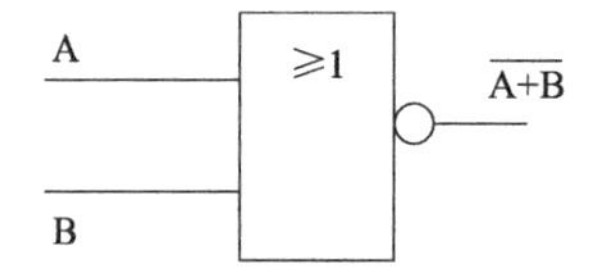

Fig. 4-5　Symbol for NOR gate

图 4-5　或非门的符号

The truth table for this logic gate is given in Table 4-5.

表 4-5 给出了该逻辑门的真值表。

Table 4-5　Truth table for the NOR gate

表 4-5　或非门的真值表

input 输入		output 输出	
A	B	OR 或	NOR 或非
0	0	0	1
0	1	1	0
1	0	1	0
1	1	1	0

4. 1. 6　XOR Gate

4. 1. 6　异或门

The XOR gate symbol is given in Fig. 4-6.

异或门的符号如图 4-6 所示。

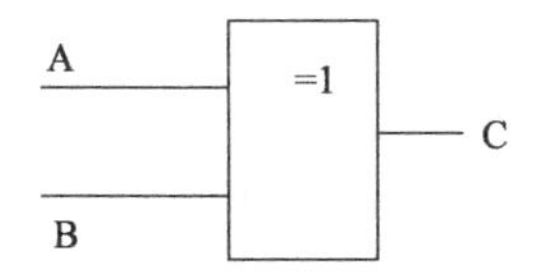

Fig. 4-6　Symbol for XOR gate
图 4-6　异或门的符号

The truth table for this logic gate is given in Table 4-6.

表 4-6 给出了该逻辑门的真值表。

Table 4-6　Truth table for the XOR gate
表 4-6　异或门的真值表

input 输入		output 输出	
A	B	OR 或	XOR 异或
0	0	0	0
0	1	1	1
1	0	1	1
1	1	1	0

The exclusive OR（XOR）is only true if the inputs A and B are opposite logic conditions. If inputs A and B are both true（1），or false（0），then the exclusive OR function is false. This is shown by the truth table above.

异或（XOR）仅在输入 A 和 B 是相反的逻辑条件时才为真，如果输入 A 和 B 都为真（1）或假（0），则异或函数为假。这种逻辑关系也可以从上面的真值表看出。

By using combinations of the various gates above we can build circuits for control. There are numbers of laws and theorems that are useful in reducing down the number of logic gates and the coding to be used for controlling a machine.

通过使用上述各种逻辑门的组合，可以构建控制电路。有许多定律和定理可用于减少逻辑门的数量和控制机器的编码数量。

4.2　Input and Output

4.2　输入和输出

PLCs mainly deal with ON/OFF inputs，such as manual switches，proximity switches，limit switches and sensors. These are used for indicating the status of a particular step，i. e.，start，stop or completion.

PLC 主要处理 ON/OFF 输入，例如手动开关、接近开关、限位开关和传感器。这些输入用于表示特定步骤的状态，即开始、停止或完成。

Outputs usually take the form of a relay，contactor，solenoid or signal lamp. The output is activated by an internal switch which takes the form of an integrated circuit.

通常用继电器、接触器、电磁阀或信号灯等作为输出。输出由内部开关激活，其中内部开关的形式是集成电路。

Input devices can take two settings in the inactivated condition, normally open (NO) and normally closed (NC), examples of which are given below in Fig. 4-7 (a). These are considered the electrical symbols.

输入设备可以在未激活状态下进行两项设置，即常开（NO）和常闭（NC）。其示例如图 4-7（a）所示，图中所示为电气符号。

Output devices are shown in Fig. 4-7 (b). Timers and counters are also available. These are very useful as it is possible to set up control sequences that are not only time dependent but also have some form of timing control or step control.

输出设备符号如图 4-7（b）所示。另外，定时器和计数器也是非常有用的，它们可以用来设置控制序列，这些控制序列不仅具有时间依赖性，而且还具有某种形式的定时控制或分级控制。

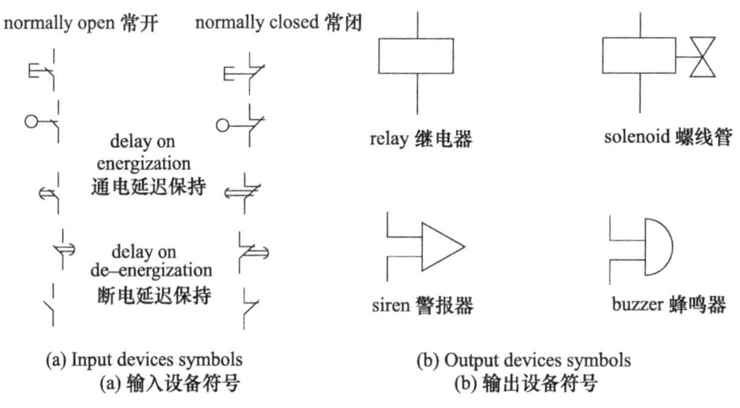

Fig. 4-7 Input/Output device symbols

图 4-7 输入/输出设备符号

4.3 Programming Language

4.3 编程语言

Since PLCs were developed to replace relays, the methods for "programming" relays have been carried over to programming PLCs. The main method is called the ladder diagram.

由于 PLC 是为了替代继电器而开发的，因此"编程"继电器的方法已经被沿用于编程 PLC，其主要方法为梯形图。

4.3.1 Ladder Diagram

4.3.1 梯形图

The ladder diagram is a pictorial representation of the required logical steps to be taken in the control process. The ladder diagram is constructed with the use of symbols to represent inputs and outputs that are connected together following the required sequence, as shown in Fig. 4-8. The inputs and outputs

梯形图是控制过程中所需要的逻辑步骤的图形化表示。梯形图由符号组成，用于表示按照所需顺序连接在一起的输入和输出，如图 4-8 所示。输入和输出排列在梯级上，而垂直线可以认为是电源

are arranged on rungs, while the vertical lines can be considered power lines, i. e. , 220V, or 24V on the left, 0V on the right, this gives rise to the form of a ladder.

线，即左边的垂直线为 220V 或 24V 的电源线，右边的垂直线为 0V 的电源线，由此而形成了梯子的形状。

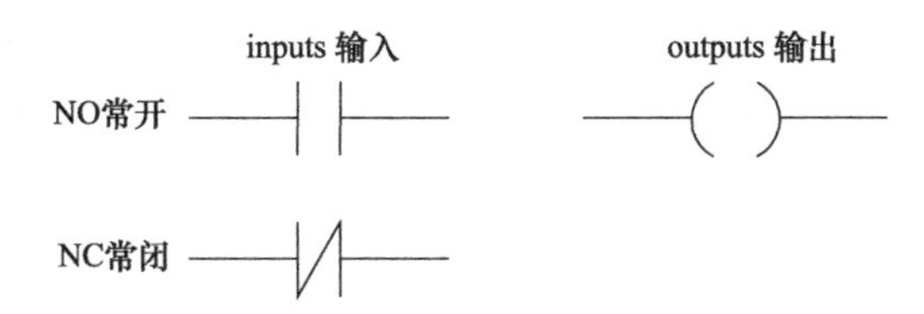

Fig. 4-8 Input and output symbols for ladder diagrams
图 4-8 梯形图的输入和输出符号

Inputs are referred to as sensor contacts and the outputs are referred to as coils (from relay coils). The state of the inputs and outputs can be considered as logic “1” or logic “0”. The CPU tests the state of the inputs for true or false conditions before setting the output to “1” or “0”, or true or false.

输入可以看做是传感器触点，输出可以看做是线圈（来自继电器线圈）。输入和输出的状态可以被认为是逻辑“1”或逻辑“0”。在将输出设置为“1”/“0”或真/假之前，CPU 会测试满足真/假条件的输入的状态。

Under the correct logical conditions, it can be considered that electrical current or power can flow from the left vertical line to the right vertical line. There are some general rules to be followed when constructing ladder diagrams:

在逻辑正确的情况下，可以认为电流或功率是从左垂直线流向右垂直线的。构建梯形图时，需要遵循以下通用规则：

① Program lines are arranged as a series of horizontal lines containing inputs (contacts) and outputs (coils).

① 程序行可以认为是包含一系列输入（触点）和输出（线圈）的水平线。

② Inputs must always precede outputs and are in the form of normally open or normally closed contacts (as shown in Fig. 4-8).

② 输入必须始终在输出之前，并采用常开或常闭触点的形式（如图 4-8 所示）。

③ There must be always one output on each ladder line. The outputs may not be connected to the left hand line, see 2 above.

③ 每条梯形线上必须始终有一个输出。输出可能未连接到左侧线，请参见第 2 条规则。

④ The numerical assignments for the inputs and outputs form part of the ladder diagram.

④ 输入和输出的数值赋值构成了梯形图的一部分。

⑤ Inputs or outputs cannot be used in a vertical manner (linking rungs).

⑤ 输入或输出不能以垂直方式表示（连接梯级）。

【Example 4-1】 It is an example of a normally open contact (IN-1) connected directly to an output (OUT-1).

【例 4-1】 直接连接到输出（OUT-1）的常开触点（IN-1）举例。

OUT-1 is ON if and only if IN-1 is ON. OUT-1 is set to logic level 1. if IN-1 is at logic level 1. If IN-1 were a switch and OUT-1 were a light bulb then the light bulb would go on and off with the operation of the switch.

当且仅当 IN-1 为 ON 时，OUT-1 为 ON。如果 IN-1 为逻辑电平 1，则 OUT-1 就为逻辑电平 1。如果 IN-1 是开关而 OUT-1 是灯泡，则灯泡将随着开关的操作而接通或断开。

Once the ladder diagram (Fig. 4-9) has been constructed then the sequence of events can be translated into an instructional list program.

一旦构建了梯形图（图 4-9），就可以将这些事件序列转换为一个程序指令列表。

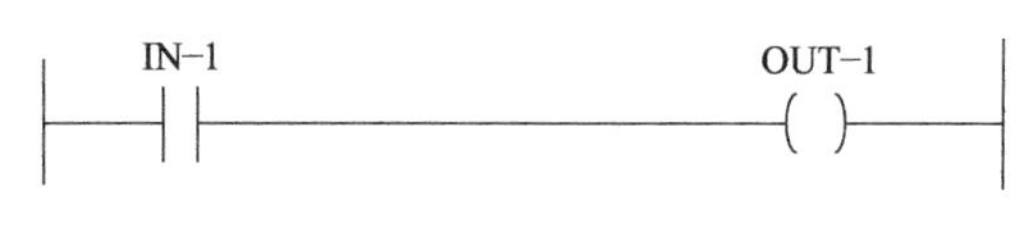

Fig. 4-9 Figure for example 4-1
图 4-9 例 4-1 附图

4. 3. 2 Instructional List Program

4. 3. 2 程序指令列表

The instructional list program consists of a series of instructions which are represented by abbreviations of the instruction they refer to e. g. , "L" or "LD" refers to "load", etc. The instruction list also contains various Boolean operations such as AND, OR, NOT, etc. By combining the various instructions, it is possible to build up the complete program as represented by the ladder diagram. The instructional list program can then be used to program the PLC. The instructional list program for the above ladder diagram could be one of Table 4-7 depending on the PLC used.

程序指令列表由一系列指令组成，这些指令由它们所引用的指令的缩写表示，例如"L"或"LD"指的是"Load"等。指令列表还包含各种布尔运算，如 AND、OR、NOT 等。通过组合各种指令，可以构建梯形图所表示的完整程序。然后可以使用指令列表程序对 PLC 进行编程。根据所使用的 PLC 的类型，上述梯形图所表示的指令列表程序可以是表 4-7 中之一。

Table 4-7 The instructional list program

表 4-7 指令列表程序

	Mitsubishi PLC 三菱 PLC	Telemecanique PLC TE 电器 PLC	Siemens PLC 西门子 PLC
Step 1 步骤 1	LD X1	L I0. 1	A I0. 1
Step 2 步骤 2	OUT Y1	ST Q0. 1	=Q0. 1
Step 3 步骤 3	EOP	END	

【Example 4-2】 Fig. 4-10 shows a slightly more complex circuit involving two input elements. IN-2 and IN-3 are both in the path from the left logic rail to OUT-2.

【例 4-2】 图 4-10 展示了包含两个输入元件的稍微复杂的电路。IN-2 和 IN-3 都位于从左侧逻辑轨道到 OUT-2 的路径中。

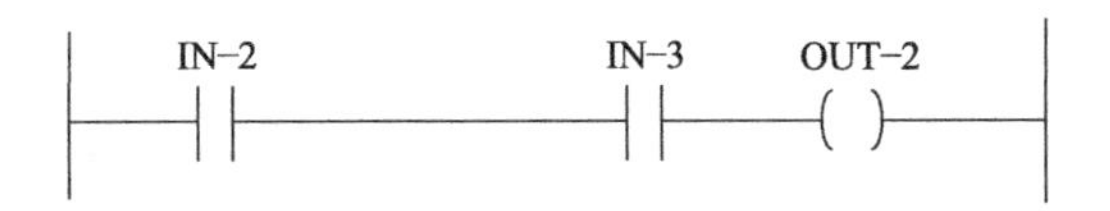

Fig. 4-10 Figure for example 4-2

图 4. 10 例 4-2 附图

OUT-2 is ON if and only if IN-2 and IN-3 are both contact ON. IN-2 "AND" IN-3 at logic level 1 means that OUT-2 will be set to logic level 1. This type of circuit is known as the logical AND connection.

当且仅当 IN-2 和 IN-3 均为 ON 时，OUT-2 为 ON。当 IN-2 "与" IN-3 为逻辑电平 1 时，那么 OUT-2 将被设置为逻辑电平 1。这种类型的电路被称为逻辑与连接。

【Example 4-3】 Fig. 4-11 shows a logical OR connection. OUT-3 will be ON if and only if IN-4 or IN-5 are contact ON. If IN-4 "OR" IN-5 is at logic level 1 then OUT-3 will be at logic level 1. This type of circuit is known as a logical OR connection.

【例 4-3】 图 4-11 所示为逻辑 OR 连接。当且仅当 IN-4 或 IN-5 为 ON 时，OUT-3 为 ON。如果 IN-4 "或" IN-5 为逻辑电平 1，则 OUT-3 将为逻辑电平 1。这种类型的电路被称为逻辑或连接。

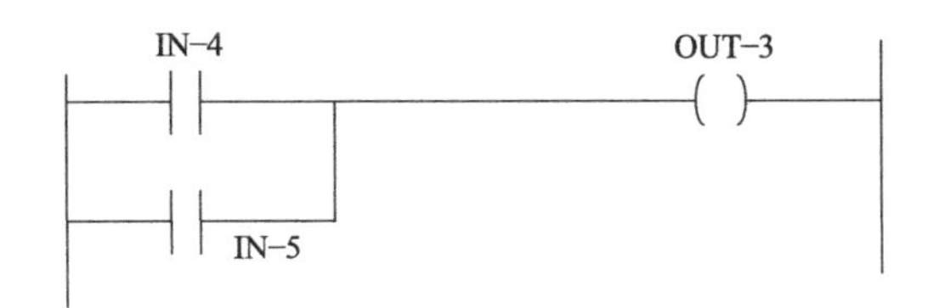

Fig. 4-11 Figure for example 4-3

图 4-11 例 4-3 附图

【Example 4-4】 Fig. 4-12 shows a circuit composed of a logical AND as well as a logical OR in the same logic rung.

【例 4-4】 图 4-12 显示了由逻辑与以及同一逻辑梯级中的逻辑或组成的电路。

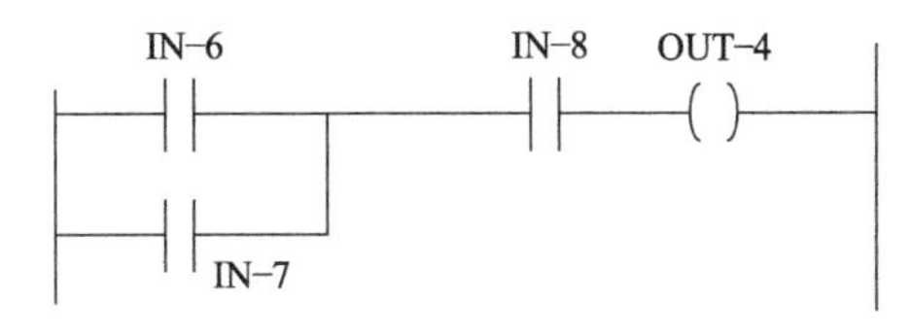

Fig. 4-12 Figure for example 4-4

图 4-12 例 4-4 附图

OUT-4 will be ON if and only if one of the following conditions is present: IN-6 or IN-7 must be contact ON, and at the same time IN-8 must also be contact ON. If either condition is not true then the output will not be ON.

如果 OUT-4 为 ON，必须满足以下条件：IN-6 或 IN-7 必须为 ON，同时 IN-8 也必须为触点 ON。如果其中一个条件不正确，则输出就不会为 ON。

【Example 4-5】 The next example to be considered is the normally closed contact. The controller does not know or care whether an input is normally open or normally closed. The controller examines the input line to determine whether it is ON or OFF regardless of its normal position. A normally closed input contact only represents the opposite condition of the input line, therefore it would be ON when the input line is not on.

【例 4-5】 接下来考虑常闭触点的例子。控制器不知道或不关心输入是常开还是常闭。不管输入的正常位置是什么，控制器都会检查输入线以确定其是 ON 还是 OFF。常闭输入触点仅表示输入线的相反状态，因此当输入线未接通时它将为 ON。

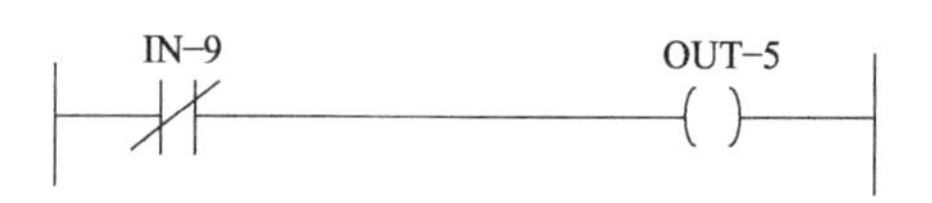

Fig. 4-13 Figure for example 4-5

图 4-13 例 4-5 附图

Fig. 4-13 shows an example of a NC contact that controls an output. OUT-5 is ON if and only if IN-9 is not ON. If this was controlling a light, the only way to get the light ON would be to leave the switch off.

图 4-13 所示为一个常闭触点控制一个输出的示例。当且仅当 IN-9 不为 ON 时，OUT-5 为 ON。如果这表示控制一个电灯，那么打开电灯的唯一方法就是关闭开关。

Another way of considering the diagram, is that the condition of the input is complete so electrical current is able to pass from left to right. An example of this would be an emergency switch, where one wants electricity to flow until the switch is activated. In logical terms, the PLC is always at the logic level “1”.

另一种考虑图 4-13 的方法是认为图中的输入条件是完整的，因此电流可以从左向右传递。这种情况的一个实例是紧急开关，人们希望电流一直流过输出直到开关被激活。在逻辑术语中，PLC 一直处于逻辑电平“1”。

4.4 Latch

4.4 锁存器

A latch will turn on a coil (relay or output) and hold it ON. The latch circuit (Fig. 4-14) is used to keep a coil energized even if the input contact which energized it subsequently turns off.

锁存器在打开线圈（继电器或输出）后会一直保持此开启状态。锁存电路（图 4-14）用于保持线圈通电，即使使其通电的输入触点随后断开。

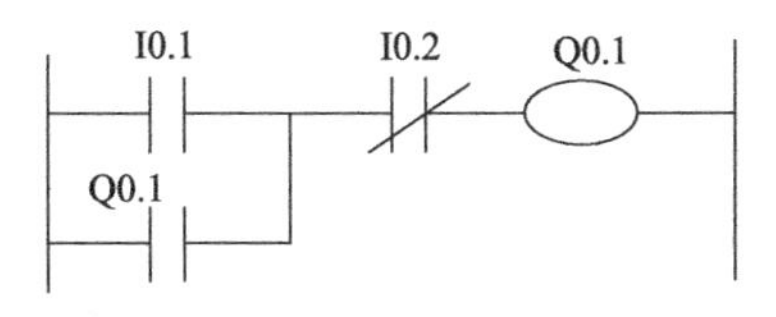

Fig. 4-14 Latch circuit
图 4-14 锁存电路

In the circuit, when input I0.1 is momentarily turned on, the output coil Q0.1 is energized or turned on. The logic state of this output is then fed back to the input (Q0.1) to keep itself energized. The latch circuit is broken or reset by momentarily turning on Input I0.2, this will switch off output Q0.1 so breaking the OR condition. Input I0.1 and Input I0.2 are usually constructed from push button switches with reset springs.

在电路中，当输入 I0.1 瞬间接通时，输出线圈 Q0.1 就会通电或接通。然后该输出的逻辑状态将反馈到输入（Q0.1）以保持自身通电。瞬时接通输入 I0.2，锁存电路会断开或复位，这将使输出 Q0.1 关闭，从而使 OR 条件中断。输入 I0.1 和输入 I0.2 通常由具有复位弹簧的按钮开关构成。

Latches can be created internally in the PLC by the use of internal relays. This can be considered an output without any external access. In the case of the Telemecanique PLC the internal relay is referred to as "Mi". This is treated and written to in the same way as any output and tested like any other input. In order to permanently set the condition of the internal relay or output, one would use the command "S". The internal relay or output can be reset to the off condition by the command "R". This is sometimes known as a flip-flop, as shown in Fig. 4-15.

可以通过使用内部继电器在 PLC 内部创建锁存器，这可以被认为是一个没有任何外部访问的输出。对于 Te 电器 PLC 来说，其内部继电器被称为"Mi"，它能够像其他输出一样以相同的方式被处理和写入，并像其他输入一样被测试。为了永久设置内部继电器或输出条件，可以使用命令"S"。可通过命令"R"将内部继电器或输出复位为关闭状态。因此其有时也被称为触发器，如图 4-15 所示。

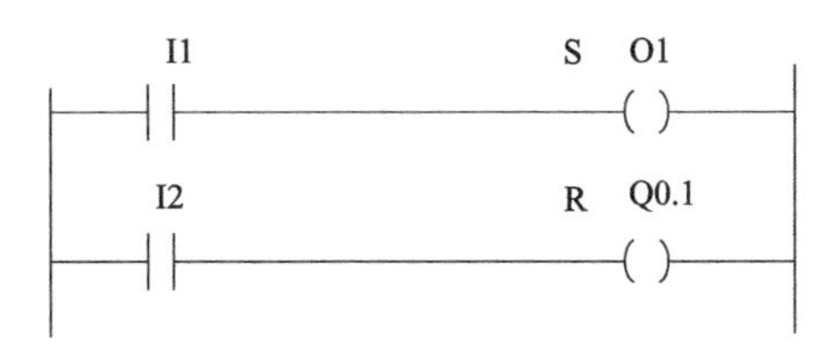

Fig. 4-15 Flip-flog
图 4-15 触发器

4.5 Counter

4.5 计数器

A PLC is equipped with counters in order to count events and make logic decisions as a result. An example would be to use a counter to count the number of parts or products going into a packaging machine. If 10 bottles per crate allowed, the PLC counts and sorts the bottles into groups of 10.

计数器是 PLC 中用于计数事件并根据计数结果做出逻辑决策的元件。例如使用计数器来计算进入包装机的零件或产品的数量，假如允许每个箱子装 10 个瓶子，那么 PLC 会计数并将瓶子分成 10 个一组。

The counter considers two values, namely the current value and the preset value. The preset value is set by the programmer when initially writing the program. The current value is the value contained in the counter every time the counter is updated.

计数器有两个值，即当前值和预设值。其中预设值由程序员在最初编写程序时设置；当前值是计数器每次更新时计数器中的值。

The current value of counter is updated in two ways either by increasing (counting up) or decreasing (counting down) the value. When the current value has reached the pre set value, the counter is said to be on. At this point all contacts that refer to this counter are activated. In this way, the counter can make logical decisions. Another function concerned with counters is the reset control. Once this has been activated then the counter will either return to the pre set value or to zero.

一般通过增加或减少数值的方式来更新计数器的当前值。当计数器的当前值达到预设值时，计数器就会打开。此时，所有与此计数器有关的触点都将被激活。通过这种方式，计数器就可以做出逻辑决策。与计数器有关的另一个功能是复位控制，一旦复位控制功能被激活，计数器将返回到预设值或零。

Fig. 4-16 shows an example where a counter is used to control a conveyor which is delivering bottles to a packing machine. Input I0. 1 starts the motor on a conveyor which latches on and stays running. Every time the bottle sensor (I0. 3) detects a bottle, it gives an input to C01 which counts up. When 10 bottles have been counted the output of the counter C01 comes on and breaks the latch. This stops the motor.

图 4-16 所示为计数器用于控制输送机的实例。此输送机可以将瓶子输送到包装机。输入 I0. 1 启动输送机上的电机，从而使系统电路被锁定并保持运转。每次瓶子传感器（I0. 3）检测到瓶子时，它都会向 C01 提供一个输入从而使 C01 对其进行计数。当计数到 10 个瓶子时，计数器 C01 的输出就会接通并中断锁存器，从而使电机停止。

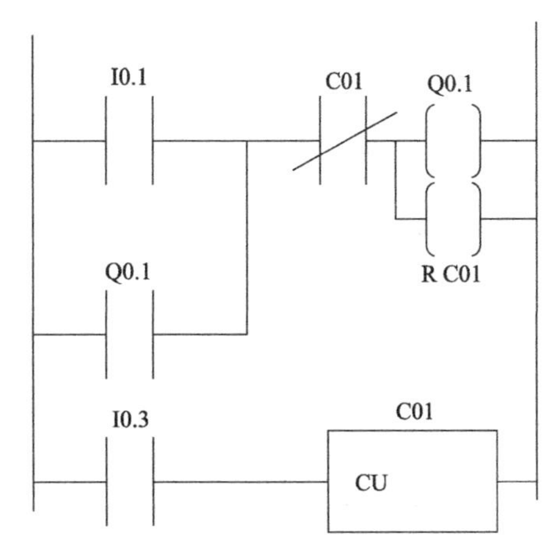

Fig. 4-16 Ladder diagram for counter
图 4-16 计数器的梯形图

4.6 Timer

A timer is the device used to keep track of and control events that need a time base. The timer has a number of values to be considered: the time base that the timer is set for, the time period or preset time required and also the current value of the timer.

The time base is set by the user or programmer. This can be a number of values, i. e., 1min, 1s, 10 ms, 100 ms.

The total time is made up of the product of the time base (units) and the time period (number of units). The time period or pre-set can be any number between 0000 and 9999.

Fig. 4-17 shows the ladder diagram and timing diagram of the timer in Telemechanique PLC. The detailed description is as follows:

- On Delay timer so Q0. 1 comes on 5s after I0. 1.
- In this case, the time period is 5.
- The input must stay on until the time delay is up.
- Command to achieve timer is IN TM1 (initiate).

4.6 定时器

定时器是用于跟踪和控制需要时基的事件的设备。定时器有许多设定值需要考虑：为定时器设置的时基、定时器需要计时的时间周期或预设时间以及定时器的当前值。

时基由用户或程序员来设置，其可以是多种值，比如1min、1s、10ms、100ms。

总时间由时基（单位）和时间周期（单位数）相乘得到，时间周期或预设值可以是0000到9999之间的任何数字。

图4-17给出了TE电器生产的PLC的定时器的梯形图和时序图，具体描述如下：

- 存在延滞定时器，所以在I0.1出现后5s Q0.1才出现。
- 时间周期为5。
- 输入必须保持打开，直到延时结束。
- 激活定时器的指令是IN TM1（开始）。

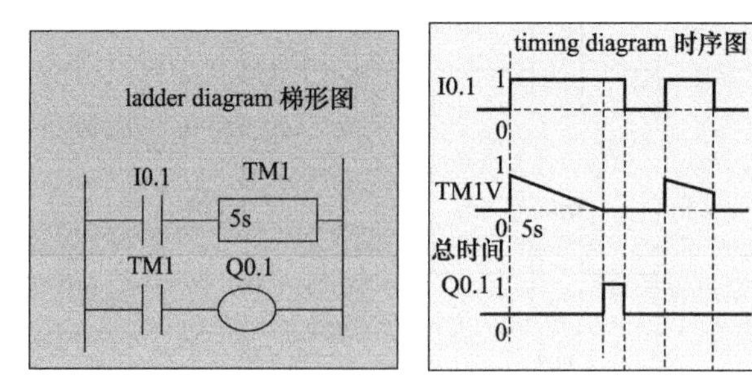

Fig. 4-17 Timer in Telemechanique PLC

图 4-17 TE 电器 PLC 中的定时器

Chapter 5 Sequential Control
第5章 顺序控制

Key words 重点词汇

sequential control 顺序控制	sequential programming 顺序程序设计
logic schematic 逻辑示意图	data command 数据指令
flow diagram 流程图	reset command 复位指令
function diagram 功能图	shift command 移位指令

5.1 The Expressions of Automation Process

5.1 自动过程的表达方式

The verbal or written description of an automatic process is usually long, difficult to follow and imprecise. The complete process is more easily understood when presented in the form of a flow diagram or function diagram. Several methods are shown in Fig. 5-1.

自动过程的口头或书面描述通常很长，并且难以理解、不够精确。当以流程图或功能图的形式来呈现自动过程时，能够使整个过程更容易理解。几种表示方式如图 5-1 所示。

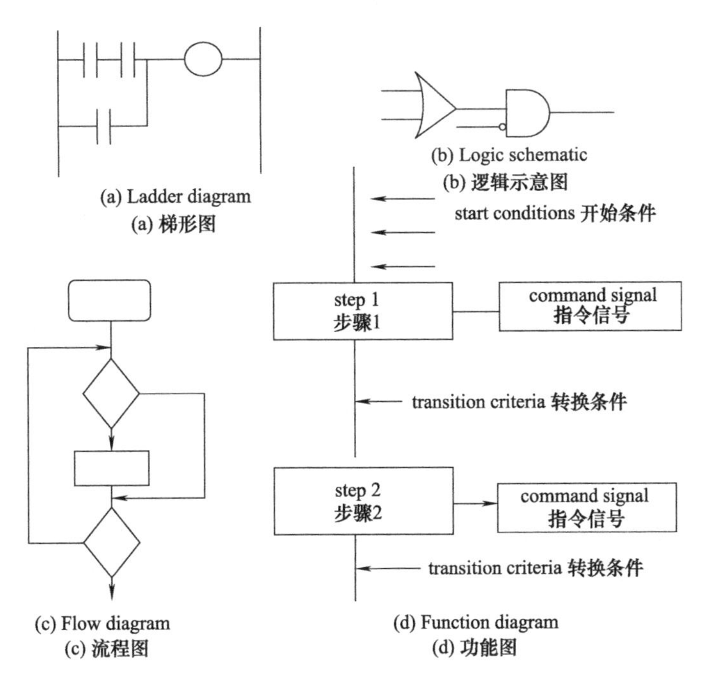

Fig. 5-1 Graphical representation of automation process
图 5-1 自动过程的图形化示意

5.1.1 Relay Ladders and Logic Schematic

Both these methods relate directly to physical circuit layouts and therefore are ideal for applications in which the PLC is replacing the conventional relay of logic system since the original drawings may be used as a basis for programming the PLC. However the layouts can be complex and very difficult to follow when used for complex sequences.

5.1.1 继电器梯形图和逻辑示意图

这两种方法都与实际的电路布局直接相关，由于原始图可以用作PLC编程的基础，因此从PLC正在取代传统逻辑系统继电器这个方面来看是理想的。然而，当用于复杂序列时，电路就会很复杂且难以理解。

5.1.2 Flow Diagram

Flow diagram more often associated with computer programming and robotics and are a common method of displaying the sequential operation of a control system. Though for larger systems it can become difficult to layout often resulting in larger cumbersome diagrams.

5.1.2 流程图

流程图通常与计算机编程有关，而且是展示控制系统顺序操作的常用方法。但是对于较大的系统，常常需要大量烦琐的示意图，从而难以对其进行布局。

5.1.3 Function Diagram

The function diagram combines most of the advantages of other methods showing the control sequence coherently and clearly. For all separate stages it shows the setting and resetting conditions, transition criteria and all necessary control signals.

5.1.3 功能图

功能图结合了其他方法的大多数优点，能够连贯、清晰地表示控制序列。对于所有不同的阶段，它能够表示出设置、重置条件，转换条件以及所有必要的控制信号。

5.2 Sequential Function Diagram

Sequential function diagram breaks a sequential task down into logical steps, the action associated with each step and the criteria to move from one step to the next (transition). They describe the required sequence in graphical format, as shown in Fig. 5-2 below.

Convention state that flows through a sequential function diagram is from top to bottom unless indicated by an arrow. The sequence is broken down into steps where actions are carried out. The transition conditions define the logical conditions that cause the sequence to move from the existing step to the next step.

5.2 顺序功能图

顺序功能图将顺序任务分解为逻辑步骤、与每个步骤相关联的动作以及从一个步骤转换到另一个步骤的条件，这些均是以图形方式来描述的，如图5-2所示。

按照相关规范，顺序功能图的流程是从上到下的，除非有箭头特别标出流程的顺序。序列被分解为执行动作的步骤（或状态）。转换条件定义了使进程从当前步骤到下一步骤的逻辑条件。

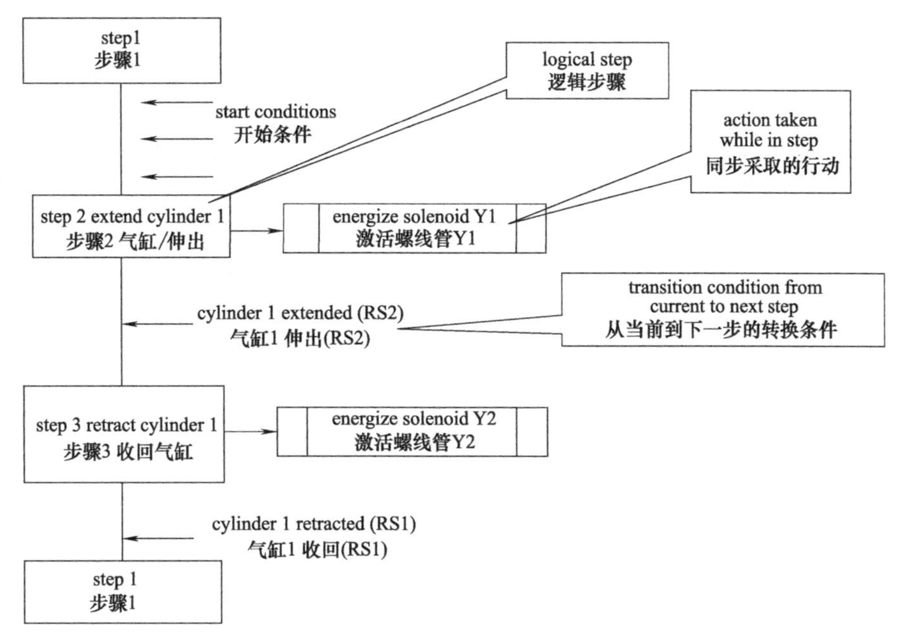

Fig. 5-2 Sequential function diagram example

图 5-2 顺序功能图示例

In the example given in Fig. 5-2, once the start conditions are met step 2 is activated. The action associated with this step is to energize solenoid Y1. When Y1 is energized the valve will switch and cylinder 1 will extend. When cylinder 1 has fully extended it will switch on reed switch 2. This is the transition condition and indicates that the action is complete. This will deactivate step 2 and activate step 3. The action associated with step 3 is to energize solenoid Y2. This will switch the valve back and cylinder 1 will retract. When cylinder 1 is fully retracted reed switch 1 will be activated, this deactivates step 3 and activates step 1 and the sequence will begin again. Actions are presented as three elements as shown in Fig. 5-3.

在图 5-2 给出的示例中，启动条件一旦被满足，则步骤 2 被激活，与该步骤相关的动作是螺线管 Y1 通电。当 Y1 通电时，阀将切换至工作位置，气缸 1 将伸出。当气缸 1 完全伸出时，簧片开关 2 闭合。这就是转换条件，表明伸出动作已完成，步骤 2 将停止并激活步骤 3。与步骤 3 相关的动作是螺线管 Y2 通电。这将使阀切换回初始工作位置，气缸 1 将收回。当气缸 1 完全收回时，簧片开关 1 将闭合，从而使步骤 3 停止并激活步骤 1，序列将再次开始。动作包含三个字段，如图 5-3 所示。

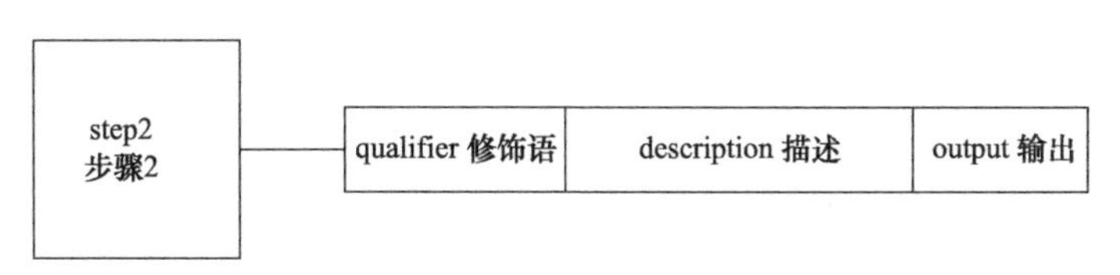

Fig. 5-3 Actions

图 5-3 动作

• A qualifier which defines what type of action, e. g. , S for set and R for reset.
• A description of the action.
• Detail of the output.

•修饰语，用于定义动作的类型，例如 S 表示 set，R 表示 reset。
•行动的描述。
•输出的详细信息。

Fig. 5-4 shows a more detailed function diagram controlling the extension and retraction of cylinder 1.

图 5-4 所示为一更详细的、控制气缸 1 伸缩的功能图。

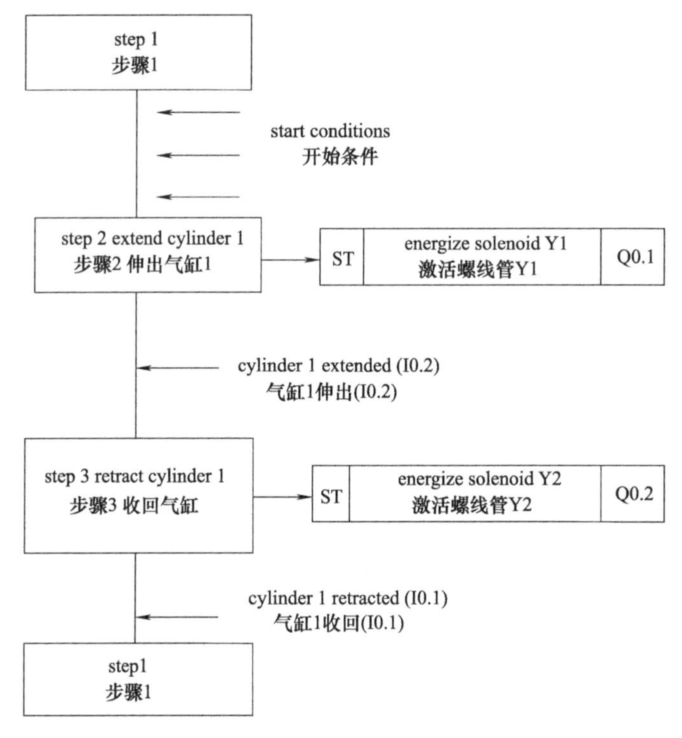

Fig. 5-4 Detailed sequential function diagram example
图 5-4 详细的顺序功能图示例

5. 3 Sequential Programming

5. 3 顺序程序设计

The program has three sections:
• Preprocessing: for input logic.
• Sequential processing: steps, transitions, transition conditions.
• Post processing: commands which control the outputs.

程序设计包括三个部分：
•预处理：针对输入逻辑。
•顺序处理：步骤、转换、转换条件。
•后处理：控制输出的命令。

In sequential processing only the steps which are active at the start of the scan and their associated instructions are executed.

在顺序处理阶段，仅执行扫描开始时有效的步骤及其相关指令。

Shift registers are one method of programming from a sequential function diagram.

使用移位寄存器，是顺序功能图编程中的一种方法。

5. 3. 1 Shift Register

5. 3. 1 移位寄存器

A register is a group of bits within the memory of the PLC. A register in which it is possible to move stored bits is called a shift register. Shift registers require at least three inputs. One is used to load data into the register, one for resetting purposes and one as a shift command. A four bit register is shown in Fig. 5-5.

寄存器是 PLC 内存中的一组位。可以移动存储位的寄存器称为移位寄存器。移位寄存器至少需要三个输入，一个用于将数据加载到寄存器中，一个用于复位，一个用作移位命令。一个四位寄存器如图 5-5 所示。

0	0	0	0	Reset: all bits go to zero 复位：所有位变为零
0	0	0	1	Data: 1st bit of register goes to logic 1 数据：寄存器的第1位变为逻辑1
0	0	1	0	Shift: the contents of the register shift one bit to the left 移位：寄存器的内容向左移1位
0	1	0	0	Shift: the contents of the register shift one bit to the left 移位：寄存器的内容向左移1位
1	0	0	0	Shift: the contents of the register shift one bit to the left 移位：寄存器的内容向左移1位
0	0	0	0	Shift: the contents of the register shift one bit to the left 移位：寄存器的内容向左移1位

Fig. 5-5 Four bit register

图 5-5 四位寄存器

5. 3. 2 Programming a Shift Register

5. 3. 2 移位寄存器的编程

The Telemecanique TSX PLC has eight shift registers (SBR0～SBR7). Each of these registers has 16 bits numbered 0 to 15. So each bit of the register is identified by a unique address in the form SBR*i*.*j*, where $i=0$ to 7 (register number) and $j=0$ to 15 (bit number).

Te 电器 TSX PLC 有 8 个移位寄存器（SBR0～SBR7）。每一个寄存器都有 16 位，编号从 0 到 15。因此，寄存器的每一个位具有 SBR*i*.*j* 形式的唯一地址标识，其中 $i=0$～7（寄存器号），$j=0$～15（位号）。

Example: for SBR0 (shift register 0), the bit addresses are shown in Fig. 5-6.

示例：SBR0（移位寄存器 0）的位地址如图 5-6 所示。

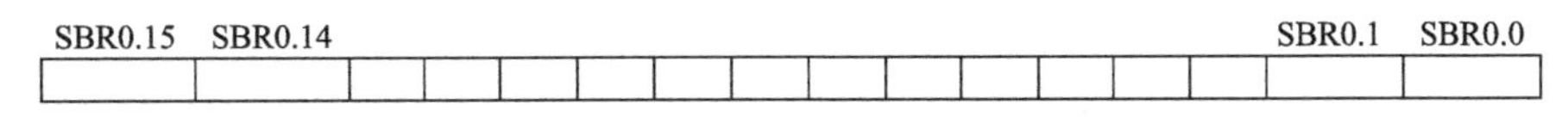

Fig. 5-6 The bit adresses of SBR0

图 5-6 SBR0 的位地址

For SBR1 the bits would be identified as SBR1.0, SBR1.1, …, SBR1.15.

对于 SBR1，这些位将被识别为 SBR1.0，SBR1.1，…，SBR1.15。

The commands used for the shift register are:

用于移位寄存器的指令有：

• Data command: set (S).
This puts logic 1 into an identified bit of the register. E.g., "S SBR0.0" will put logic 1 into bit SBR0.0.

•数据指令：设置（S）。
此指令将逻辑 1 置于寄存器已识别的位中。例如"S SBR0.0"，即将逻辑 1 置于 SBR0.0 位。

• Reset command: reset (R).
This will reset an identified bit to logic 0. E.g., "R SBR0.0" will put logic 0 into SBR0.0.
"R SBR0" resets the entire register by putting logic 0 into each bit.

•复位指令：复位（R）。
此指令会将已识别的位复位为逻辑 0。例如"R SBR0.0"，即将逻辑 0 置入 SBR0.0；"R SBR0"通过将逻辑 0 置于每个位来复位整个寄存器。

• Shift command: CU/CD.
This shifts the contents of the regiser one bit to the left/right. E.g., "CU SBR0" will shift all the bits in the register SBR0 to the left. "CD SBR0" will shift all the bits in the register to the right.

•移位指令：CU/CD。
此指令将寄存器的内容向左/右移一位。例如"CU SBR0"即将寄存器 SBR0 中的所有位向左移一位；"CD SBR0"即将寄存器中的所有位向右移一位。

5.4 Example of Shift Register Programming

5.4 移位寄存器编程实例

A PLC is used to control a light. When the switch is pressed the light should come on for 1 second then turn off. The address list is shown in Table 5-1.

PLC 用于控制一个指示灯，当按下开关时，灯亮 1 秒钟然后熄灭，此过程的地址列表见表 5-1。

Table 5-1 Address list
表 5-1 地址列表

inputs 输入		outputs 输出	
switch 开关	I0.1	light 1 灯 1	Q0.1

Fig. 5-7 shows the sequential function diagram for this sequence. While Table 5-2 illustrates what is happening in shift register SBR0 at each step in the sequence.

图 5-7 为此序列的顺序功能图，表 5-2 给出了移位寄存器 SBR0 在序列的每个步骤中的状态。

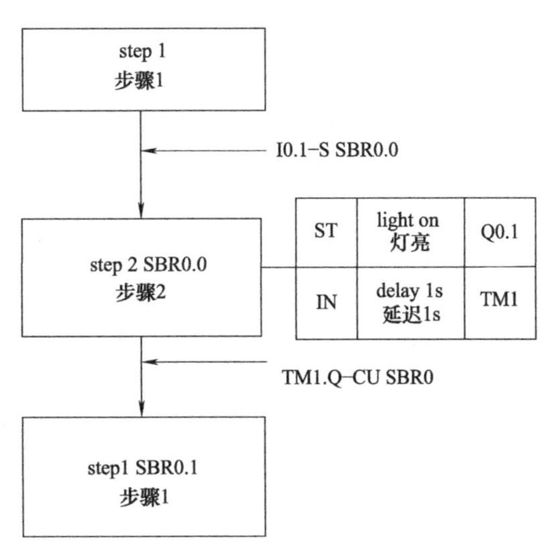

Fig. 5-7 Sequential function diagram

图 5-7 顺序功能图

Table 5-2 SBR0 at each step in the sequence

表 5-2 序列中每个步骤的 SBR0

output 输出	SBR0. 15	…	SBR0. 4	SBR0. 3	SBR0. 2	SBR0. 1	SBR0. 0	input 输入
Q0. 1, TM1	0	…	0	0	0	0	1	I0. 1
	0	…	0	0	0	1	0	TM1. Q
Q0. 1, TM1	0	…	0	0	0	1	1	I0. 1
	0	…	0	0	1	1	0	TM1. Q
Q0. 1	0	…	0	0	1	1	1	I0. 1

When I0. 1 is ON a set command is given to the shift register (S SBR0. 0). This will put a logic 1 into the first bit of the register (SBR0. 0). This activates step 2 of the sequence. This will turn ON Q0. 1 so light 1 will come on. It will also start the timer TM1. The timer is programmed to switch ON output TM1. Q after 1s.

当 I0. 1 为 ON 时，给移位寄存器发出一个数据指令（S SBR0. 0），从而使寄存器的第一位被置为逻辑 1（SBR0. 0）。以上过程激活序列的第 2 步，从而使 Q0. 1 变为 ON，因此 Light 1 亮起，同时定时器 TM1 启动。定时器被编程为在 1s 后使输出 TM1. Q 变为 ON。

TM1. Q comes ON and gives a shift command (CU SBR0) to the register so the logic 1 shifts from the first bit of the register to the second bit (SBR0. 1). This deactivates step 2 so Q0. 1 (the light) will go off.

当 TM1. Q 为 ON 后，会向寄存器发出移位指令（CU SBR0），因此逻辑 1 从寄存器的第一位移到第二位（SBR0. 1）。这将使步骤 2 停止并再次激活步骤 1，即 Q0. 1（指示灯）将会熄灭。

The sequence will stop until I0. 1 is switched ON again and a set command is given to the register, a logic 1 goes into SBR0. 0, Q0. 1 goes ON and the timer starts.

当 I0. 1 再次为 ON 时，序列会停止，此时会向寄存器发出设置指令，SBR0. 0 变为逻辑 1，Q0. 1 为 ON，定时器启动。

Fig. 5-8 shows the ladder diagram for this sequence. The sequential processing stage includes the first two rungs of the ladder. The first rung of the ladder shows that I0. 1 will set the register and start the sequence. The second rung of the ladder shows the conditions for shifting the register.

图 5-8 为该序列的梯形图。顺序处理阶段包括梯子的前两个梯级。梯形图的第一个梯级表示 I0. 1 将设置寄存器并启动序列。梯子的第二个梯级表明了移位寄存器发生移位的条件。

The third rung of the ladder is the post processing stage and shows that SBR0. 0 switches on Q0. 1 (the light) and starts the timer.

梯形图的第三个梯级是后处理阶段，其表示 SBR0. 0 打开 Q0. 1（指示灯）并启动计时器。

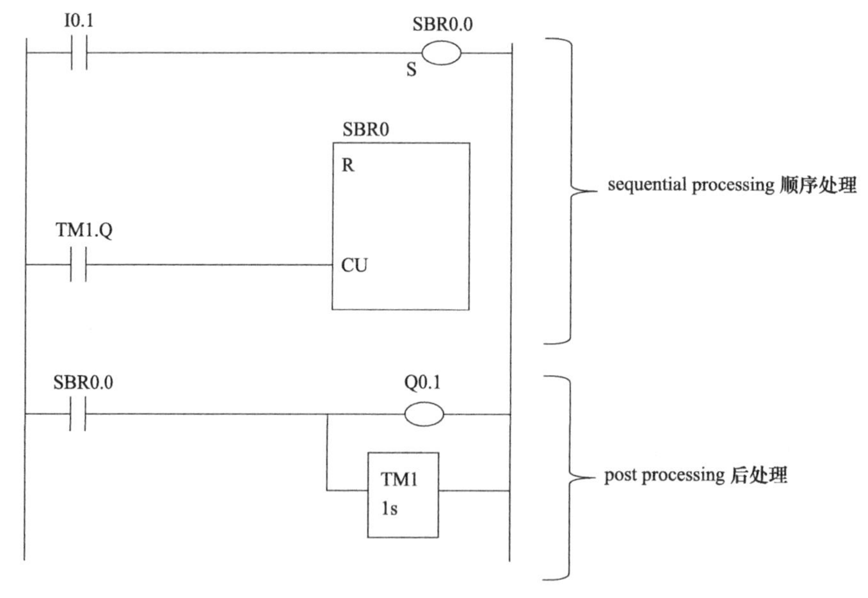

Fig. 5-8 Ladder diagram
图 5-8 梯形图

Chapter 6　PID Control
第6章　PID控制

Key words 重点词汇

proportional control　比例控制	phase margin　相位裕量
integral control　积分控制	amplitude margin　幅值裕量
differential control　微分控制	frequency characteristic　频率特性
open-loop transfer function　开环传递函数	parameter tuning　参数整定
closed-loop transfer function　闭环传递函数	step response　阶跃响应

6.1　Introduction of PID Control

In practical engineering, the most widely used control laws for controllers are proportional, integral, and derivative control, which can be called PID control, also known as PID regulation. These controllers are connected in series in the forward channel of the system, thus playing the role of series correction.

PID controller has been nearly 80 years since it comes out. It has become one of the main technologies of industrial control due to its simple structure, good stability, reliable work and convenient adjustment. When the structure and parameters of the controlled object can not be fully grasped, or an accurate mathematical model can not be obtained, and other techniques of control theory are difficult to adopt, the structure and parameters of the controller must be determined by experience and on-site debugging, thus using PID control technology is the most convenient. The PID controller is based on the error of the system, using proportional, integral and differential operations to calculate the control amount. Typical PID control block diagram is shown in Fig. 6-1.

6.1　PID控制简介

在工程实际中，应用最为广泛的控制器控制规律为比例、积分、微分控制，简称PID控制，又称PID调节。这些控制器是串接在系统前向通道中的，因而起着串联校正的作用。

PID控制器问世至今已有近80年的历史，因其结构简单、稳定性好、工作可靠、修整方便而成为工业控制的主要技术之一。当被控对象的结构和参数不能完全掌握，或得不到精确的数学模型时，控制理论的其他技术难以使用，系统控制器的结构和参数必须依靠经验和现场调试来确定，此时应用PID控制技术最为方便。PID控制器就是根据系统的误差，利用比例、积分、微分计算出控制量进行控制的装置。典型PID控制框图如图6-1所示。

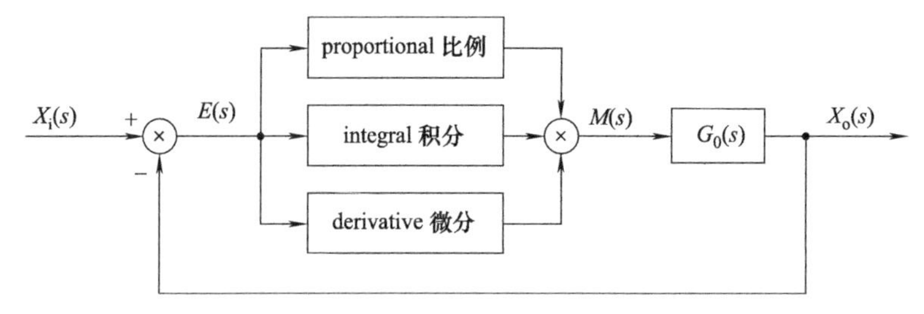

Fig. 6-1 PID control block diagram

图 6-1 PID 控制框图

For the continuous control system, PID control law is expressed as

在连续控制系统中采用的 PID 控制规律为

$$u(t)=K_{P}e(t)+K_{I}\int_{0}^{i}e(t)\mathrm{d}t+K_{D}\frac{\mathrm{d}e(t)}{\mathrm{d}t} \tag{6-1}$$

where $u(t)$ is the control amount; K_P is the proportional gain; $e(t)$ is the system control deviation; K_I is the integral time constant; K_D is the derivative time constant.

式中，$u(t)$ 为控制量；K_P 为比例增益；$e(t)$ 为系统的控制偏差；K_I 为积分时间常数；K_D 为微分时间常数。

The output of proportional control is proportional to the current deviation signal $e(t)$. The principle is simple. The larger the K_P, the better the system dynamics. But too large K_P will cause system oscillations, reduce system stability, and at the same time can not remove system steady-state errors.

比例控制的输出与当前时刻的偏差信号 $e(t)$ 成比例，原理简单，K_P 越大，系统动态性越好，但 K_P 太大会引起系统振荡，降低系统稳定性，同时也不能去除系统的稳态误差。

Integral control is introduced to eliminate the steady-state error of the system. The integral term increases with time, so that the controller reduces the steady-state error until the error is eliminated. But the influence of system disturbance can not be overcome in time.

积分控制是为了消除系统的稳态误差而引入的，积分项随着时间的增加而增大，从而使控制器减小稳态误差，直到误差被消除，但不能及时克服系统扰动的影响。

Derivative control can reflect the change speed of $e(t)$. When the deviation appears, it has a great control effect. It has the function of advanced control, helps reduce overshoot and adjustment time, and improves the dynamic characteristics of the system, but it can not eliminate the steady-state error of the system.

微分控制可以反映 $e(t)$ 的变化速度，在偏差刚出现时产生很大的控制作用，具有超前控制的能力，有助于减小超调和调整时间，改善系统的动态特性，但其不能消除系统的稳态误差。

Since computer control adopts sampling control, the control quantity can only be calculated based on the sampling deviation. Therefore, it is necessary to discretize the differential equations of continuous systems and replace them with numerical difference equations. Discrete way: define T_s as the sampling period, then the deviation is $e(k)$ at the kT_s moment. The integral part is expressed in the form of addition and sum, i. e. , $e(k)+e(k+1)+\cdots$. The differential link is expressed in the form of slope, i. e. , $[e(k)-e(k-1)]/T_s$. The following rules can be obtained.

由于计算机控制采用采样控制，只能根据采样偏差计算控制量，因此需要对连续系统的微分方程进行离散化处理，将其用数字形式的差分方程代替。离散化方式为：采样周期为 T_s，则在第 kT_s 时刻：偏差 $e(k)$；积分环节用加和的形式表示，即 $e(k)+e(k+1)+\cdots$；微分环节用斜率的形式表示，即 $[e(k)-e(k-1)]/T_s$，可以得到以下规律。

The control law of the position digital PID is

位置型数字 PID 控制规律为

$$u(k)=K_Pe(k)+K_I\sum_{n=0}^{k}e(n)T_s+K_D\frac{e(k)-e(k-1)}{T_s} \tag{6-2}$$

The control law of the incremental digital PID is

增量型数字 PID 控制规律为

$$\Delta u(k)=K_P[e(k)-e(k-1)]+K_IT_se(k)+K\frac{K_D}{T_s}[e(k)-2e(k-1)+e(k-2)] \tag{6-3}$$

where k is a sample number; $u(k)$ is the control quantity at sampling time k.

式中，k 为采样序号；$u(k)$ 为采样时刻 k 时的控制量。

The above is the incremental representation of the discrete PID. It can be seen from the formula that the incremental expression result is related to the deviation of the last three times, which greatly improves the stability of the system. In many practical situations, the PID control does not necessarily require all the three control action. According to different combinations of the proportional control, integral control and derivative control, there are generally four types of PID controllers including P, PI, PD and PID control.

以上就是离散化 PID 的增量式表示方式。由公式可以看出，增量式的表达结果和最近三次的偏差有关，这样就大大提高了系统的稳定性。实际在很多情形下，PID 控制并不一定需要全部的三种控制作用。根据比例控制项、积分控制项和微分控制项的不同组合，PID 控制器通常存在 P、PI、PD 和 PID 控制四种形式。

6.2 Proportional Control

6.2 比例控制

Proportional (P) control is the simplest control method. The output of the controller is proportional to the input error signal, and its transfer function is

比例（P）控制是一种最简单的控制方式。其控制器的输出与输入误差信号成比例关系，其传递函数为

$$G_c(s)=\frac{M(s)}{E(s)}=K_P \tag{6-4}$$

The output signal of the P control is

P控制的输出信号为

$$u(t)=K_P e(t) \tag{6-5}$$

where K_P is proportional gain, also known as proportional amplification factor.

式中，K_P 为比例增益，也称为比例放大倍数。

Proportional control only changes the gain of the system without affecting the phase. It's influence on the system is mainly reflected in the steady-state error and stability of the system. Increasing the proportional gain can improve the open-loop gain of the system and reduce the steady-state error of the system, so as to improve the control accuracy of the system. However, it will reduce the relative stability of the system and even cause the instability of the closed-loop system.

比例控制只改变系统的增益而不影响相位，它对系统的影响主要反映在系统的稳态误差和稳定性上。增大比例增益可提高系统的开环增益，减小系统的稳态误差，从而提高系统的控制精度。但这会降低系统的相对稳定性，甚至可能造成闭环系统的不稳定。

Fig. 6-2 shows a certain direct current (DC) motor position control system. θ_i is the input angle through potentiometer, θ_o is the output angle, K_s is a potentiometer sensitivity (V/rad), K_P is the amplification factor, K_m is a motor constant (N・m/V), J is the motor rotor moment of inertia (kg/m^2), F is an equivalent viscous friction coefficient (N・m・s/rad), n is the reduction ratio of the gearbox.

图6-2所示为某直流电机位置控制系统。θ_i 为通过电位计输入的角度，θ_o 为电机输出转角，K_s 为电位计的灵敏度（V/rad），K_P 为放大倍数，K_m 为电机常数（N・m/V），J 为电机转子转动惯量（kg/m^2），F 是等效黏滞摩擦系数（N・m・s/rad），n 为减速箱的减速比。

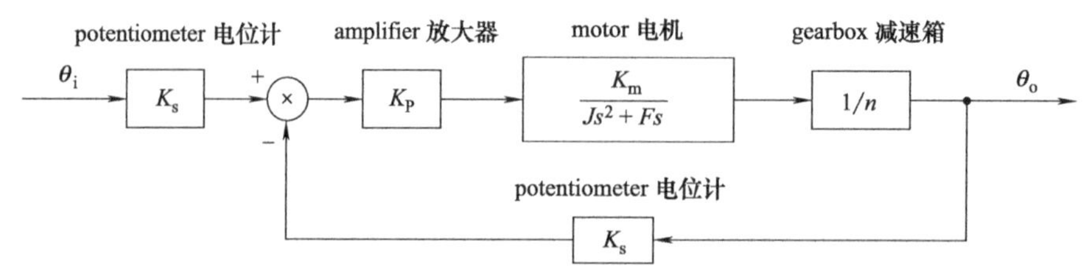

Fig. 6-2 Motor position control system

图6-2 电机位置控制系统

According to Fig. 6-2, the open-loop transfer function of the system is

由图6-2可知系统的开环传递函数为

$$G_K(s)=\frac{K_sK_PK_m}{ns(Js+F)}=\frac{K_sK_PK_m/nF}{s(\frac{J}{F}s+1)} \tag{6-6}$$

According to the open-loop transfer function, we can get its open-loop gain as

根据开环传递函数可知，其开环增益为

$$K=K_sK_PK_m/nF$$

When the system is unloaded and the input signal has a constant speed, the steady-state error is

当系统空载且输入恒速度信号时，系统稳态误差为

$$e_{ss}=\frac{1}{K}=\frac{nF}{K_sK_PK_m}=\theta_o-\theta_i \tag{6-7}$$

Therefore, K_P can be increased to increase the open-loop gain and reduce the steady-state error. At the same time, the analysis shows that the steady-state error is a constant value after K_P is set. Therefore, the error compensation value e_{ss} can be increased by the input signal θ_i to achieve the function of eliminating the error of the output signal. However, due to the fact that there is always a load in the actual system, which is equivalent to the dynamic change of the moment of inertia J and viscous friction coefficient F of the motor rotor, resulting in the steady-state error e_{ss} dynamically changing with the working conditions. In this case, the error compensation function is difficult to achieve.

可以通过增大 K_P，提高开环增益，减小稳态误差。同时，分析可知，当 K_P 调定后，稳态误差是个恒定值，因此可通过输入信号 θ_i 增加误差补偿值 e_{ss}，以达到消除输出信号误差的目的。但是，由于实际系统总是存在负载，并且随工况动态变化，也就是说电机转子转动惯量 J 和黏滞摩擦系数 F 动态变化，导致稳态误差 e_{ss} 随工况动态变化。这种情况下，误差补偿就难以实现。

Therefore, when there is only proportional control, the system output always has a steady-state error. According to the previous study, in order to improve the steady-state performance index of the system and reduce the steady-state error of the system, a feasible way is to increase the steady-state error coefficient of the system, that is to increase the open-loop gain of the system. But only when $K_P\to\infty$, the output of the system can track the input, which will inevitably destroy the dynamic performance and stability of the system.

因此，当仅有比例控制时，系统输出总是存在稳态误差。根据前面所学，为了提高系统的稳态性，减少系统的稳态误差，一个可行的办法是提高系统的稳态误差系数，即增加系统的开环增益。但只有当 $K_P\to\infty$时，系统的输出才能跟踪输入，而这必将破坏系统的动态性能和稳定性。

6.3 Proportional-integral Control

6.3 比例-积分控制

For an automatic control system, if there is a steady-state error after entering the steady-state, the control system is said to have a steady-state error or it is abbreviated as a differential system. In order to eliminate

对一个自动控制系统，如果在进入稳态后存在稳态误差，则称这个控制系统是有稳态误差的或简称有差系统。为了消除稳态误差，

the steady-state error, the "integral term" must be introduced into the controller. The effect of integral term on error depends on the integration of time. As time increases, the integral term will increase, even if the error is very small, the integral term will also increase with time. It makes the output of the controller increase, and then the steady-state error further decreases until it is equal to zero. However, due to the phase lag introduced by the integral term, the stability of the system becomes worse. Therefore, the integral control is generally not used alone. It is usually combined with proportional control to form the proportional-integral (PI) control whose transfer function is

在控制器中必须引入“积分项”。积分项对误差的作用取决于时间的积分。随着时间的增加，积分项会增大，即便误差很小，积分项也会随着时间的增加而增大，它推动控制器的输出增大进而使稳态误差进一步减小，直到等于零。但是由于积分引入了相位滞后，使系统稳定性变差，因此积分控制一般不单独使用，而是与比例控制结合构成比例-积分（PI）控制，其传递函数为

$$G_c(s)=\frac{M(s)}{E(s)}=K_P\left(1+\frac{1}{T_I s}\right) \tag{6-8}$$

The output of the controller is

控制器的输出为

$$u(t)=K_P\left[e(t)+\frac{1}{T_I}\int_0^t e(t)\mathrm{d}t\right] \tag{6-9}$$

where T_I is the integral time constant.

式中，T_I 为积分时间常数。

The logarithmic frequency characteristic curve is plotted in Fig. 6-3. Seen from this figure, PI correction can improve the type order of the system and eliminate or reduce the steady-state error. However, the system phase lags behind, the phase margin decreases and the relative stability deteriorates. Therefore, PI control is usually used when the stability margin is large enough.

绘制其对数频率特性曲线，如图 6-3 所示。由图 6-3 可知，PI 校正使系统型次提高，稳态误差得以消除或减小；但是系统相位滞后，相位裕度减低，相对稳定性变差。因此，PI 控制一般在稳定裕度足够大时采用。

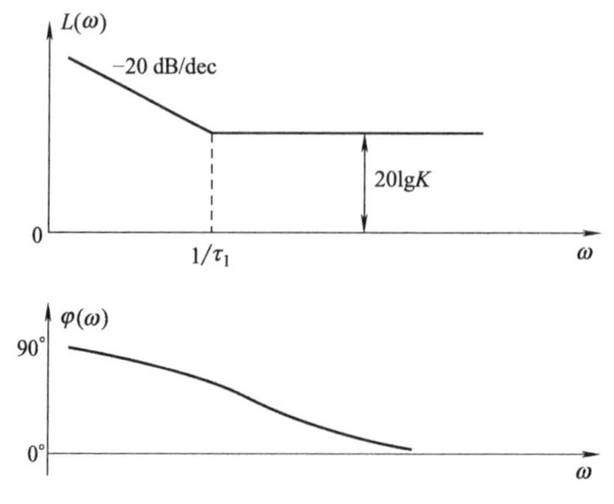

Fig. 6-3 Logarithmic frequency characteristic of PI correction part

图 6-3 PI 校正环节对数频率特性曲线

Still take the motor position control system as an example, as shown in Fig. 6-4.

仍以电机位置控制系统为例，如图 6-4 所示。

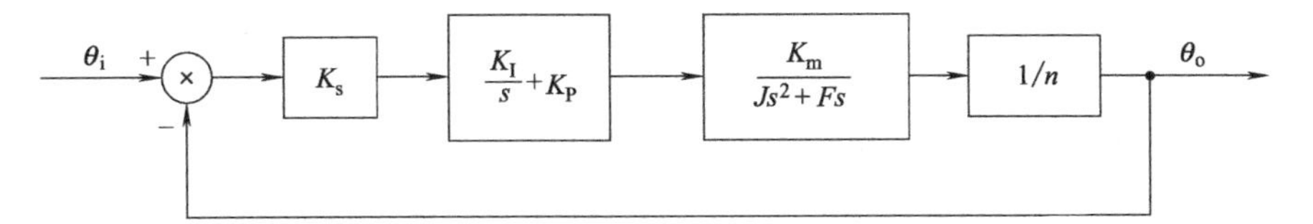

Fig. 6-4 System block diagram of a motor position PI control

图 6-4 电机位置 PI 控制系统框图

From the Fig. 6-4, the open-loop transfer function of the system is achieved as

由图 6-4 可知，系统开环传递函数为

$$G_K(s)=\frac{\dfrac{K_sK_PK_mK_I}{nF}\left(\dfrac{K_P}{K_I}s+1\right)}{s^2\left(\dfrac{J}{F}s+1\right)} \tag{6-10}$$

According to the open-loop transfer function, PI correction eliminates or reduces the steady-state error. But the relative stability is reduced, and even if K_P and K_I are unreasonably selected, the system may become unstable. A typical PI controller is shown in Fig. 6-5.

根据其开环传递函数可知，PI 校正使系统稳态误差得以消除或减小，但是相对稳定性降低，如果 K_P 和 K_I 选择不合理，有可能造成系统不稳定。典型 PI 控制器如图 6-5 所示。

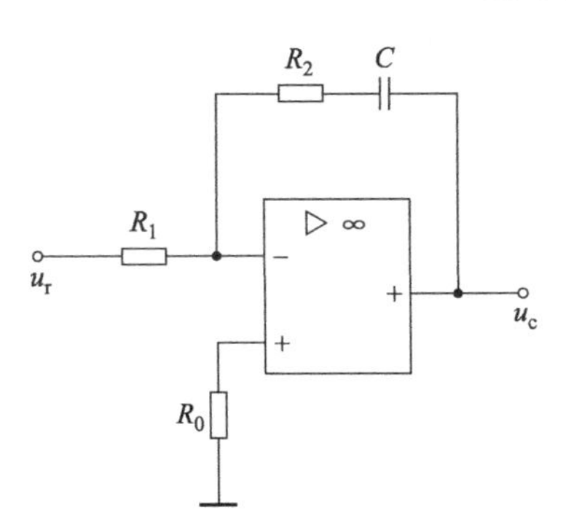

Fig. 6-5 A typical PI controller

图 6-5 典型 PI 控制器

6.4 Proportional-derivative Control

6.4 比例-微分控制

During the adjustment process to overcome the error, the automatic control system may oscillate or even lose stability. The reason is that there are large inertial components (links) or hysteresis components, which have the effect of suppressing errors, and their changes

自动控制系统在克服误差的调节过程中可能会出现振荡甚至失稳，其原因是存在较大惯性组件（环节）或滞后组件。这些组件具有抑制误差的作用，其变化总是落后

always lag behind the changes in errors. The solution is to make the change of the effect of the error suppression "leading", that is, when the error closes to zero, the effect of suppressing the error should be zero. This requires increasing the "derivative term".

于误差的变化。解决的办法是使抑制误差的作用的变化“超前”，即在误差接近零时，抑制误差的作用就应该是零。这就需要增加“微分项”。

Differential control whose input is proportional to the differential of the output error signal (i. e. , the rate of error change) can predict the change trend of the error. In this way, the differential controller can make the control effect of restraining error equal to zero or even negative value in advance, so as to avoid serious overshoot of controlled variables. Because the differential control reflects the rate of change of the error, only when the error changes with time, the differential control will have an effect on the system, and will not work on objects that are unchanged or slowly changing. Therefore, the derivative control can not be used alone in series with the controlled object under any circumstances. It is usually combined with proportional control to form the proportional -derivative (PD) control, and it's transfer function is

微分控制的输入与输出误差信号的微分（即误差的变化率）成正比关系，能预测误差变化的趋势。这样，微分控制器就能够提前使抑制误差的控制作用等于零，甚至为负值，从而避免了被控量的严重超调。但是由于微分控制反映误差的变化率，只有当误差随时间变化时，微分控制才会对系统起作用，而对无变化或缓慢变化的对象，则不起作用。因此微分控制在任何情况下不能单独与被控制对象串联使用，通常结合比例控制构成比例-微分（PD）控制，其传递函数为

$$G_c(s)=\frac{M(s)}{E(s)}=K_P(1+T_D s) \tag{6-11}$$

The output of the PD controller is

PD 控制器的输出为

$$u(t)=K_P\left[e(t)+T_D\frac{de(t)}{dt}\right] \tag{6-12}$$

where T_D is the derivative time constant.

式中，T_D 为微分时间常数。

The logarithmic frequency characteristic curve is plotted in Fig. 6-6. Obviously, the PD correction makes the phase lead, increases the system phase margin, enhances the stability, and increases the amplitude crossover frequency and the response speed. Therefore, for the controlled object with greater inertia or lag, the PD controller can improve the dynamic characteristics of the system during the adjustment process. However, the high frequency gain increases and the disturbance rejection ability weakens. A typical PD controller is shown in Fig. 6-7.

绘制其对数频率特性曲线，如图 6-6 所示。显然 PD 校正使相位超前，增加系统相位裕量，增强稳定性，并且幅值穿越频率增加，响应速度提高。所以对有较大惯性或滞后的被控对象，PD 控制器能改善系统在调节过程中的动态特性，但是高频增益上升，系统抗干扰能力减弱。典型 PD 控制器如图 6-7 所示。

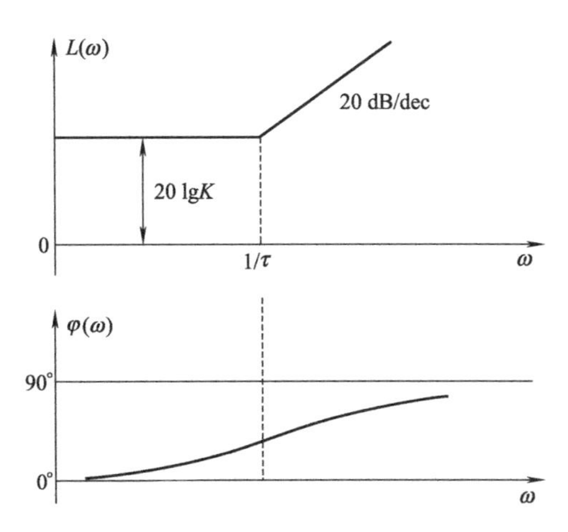

Fig. 6-6 Logarithmic correction frequency characteristic of PD correction part
图 6-6 PD 校正环节对数频率特性曲线图

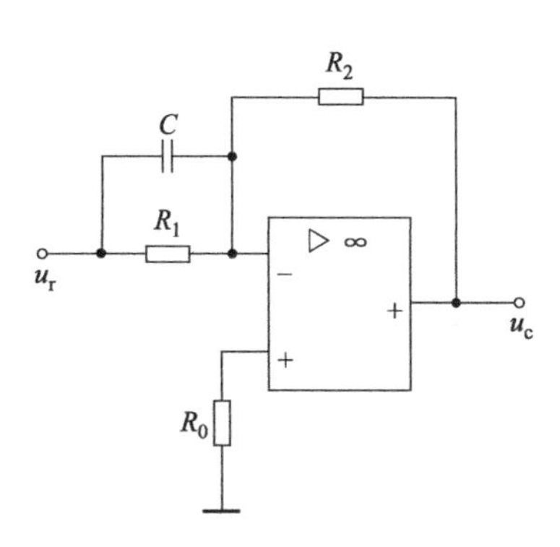

Fig. 6-7 A typical PD controller
图 6-7 典型 PD 控制器

6.5 Proportional-integral-derivative Control

6.5 比例-积分-微分控制

Proportional-integral-derivative (PID) controller is superimposed by proportional, integral, and derivative control actions. It's transfer function is

比例-积分-微分（PID）控制器是比例、积分、微分三种控制作用的叠加，其传递函数为

$$G_c(s)=\frac{M(s)}{E(s)}=K_P\left(1+\frac{1}{T_I s}+T_D s\right) \tag{6-13}$$

The output of the PID controller is

PID 控制器的输出为

$$u(t)=K_P\left[e(t)+\frac{1}{T_I}\int_0^t e(t)\mathrm{d}t+T_D\frac{\mathrm{d}e(t)}{\mathrm{d}t}\right] \tag{6-14}$$

PID control eliminates errors through integral action, reduces overshoot and speeds up response through derivative action. It is a control that combines the advantages of PI control and PD control and removes their shortcomings. From the perspective of frequency domain, it's logarithmic frequency characteristic curve can be draw as Fig. 6-8. It can be seen from the figure that in the low -frequency band, the integral effect is used to improve the steady-state accuracy of the system; in the middle-frequency band, the slope is kept unchanged, and the proportional function is used to increase the system shear frequency, occupy a wider frequency band and improve the system response speed; in the high-frequency band, the derivative effect is used to increase the phase margin and improve the system stability.

PID 控制通过积分作用消除误差，通过微分作用缩小超调量，加快响应速度，是综合了 PI 控制与 PD 控制的控制。从频域角度，绘制其对数频率特性曲线，如图 6-8 所示。由图 6-8 可知，在低频段，利用积分效应提高系统稳态精度；在中频段保持斜率不变，并利用比例作用增大系统剪切频率，占据较宽的频带，提高系统响应速度；在高频段，利用微分效应，增加相位裕度，提高系统稳定性。

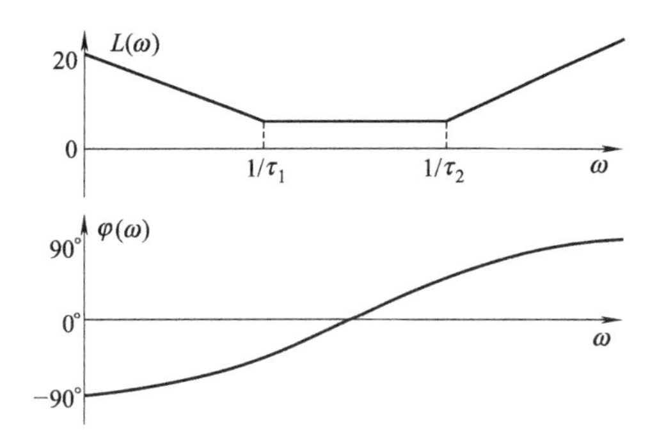

Fig. 6-8 Logarithmic graph showing the frequency characteristic of PID correction part

图 6-8 PID 校正部分的对数频率特性曲线

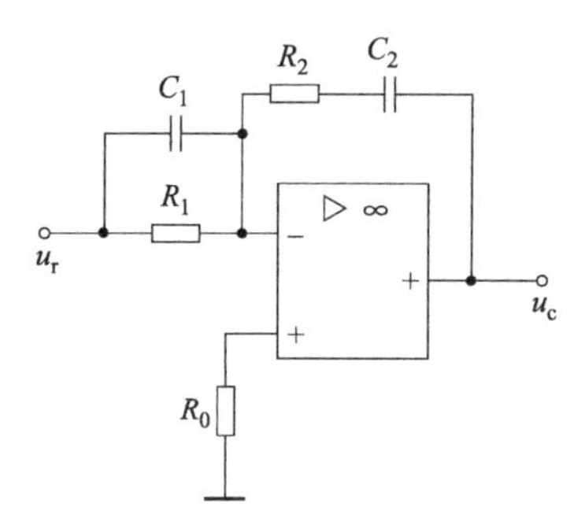

Fig. 6-9 A typical PID controller

图 6-9 典型的 PID 控制器

A typical PID controller is shown in Fig. 6-9.

典型 PID 控制器如图 6-9 所示。

In summary, we can see that PID control is the process of controlling deviation. If the deviation is zero, the proportional link does not work. Only when there is deviation, the proportion link works. The integral link is mainly used to eliminate the static error which is the difference between the output value and the set value after the system is stable. The integral link is actually the process of error accumulation, which adds the accumulated error to the original system to offset the static error caused by the system. The derivative link reflects the variation law or the change trend of the deviation signal. According to the change trend of the deviation signal, we can make advance adjustment, thus increasing the rapidity of the system.

综上所述，PID 控制就是对偏差进行控制的过程。如果偏差为 0，则比例环节不起作用；只有存在偏差时，比例环节才起作用。积分环节主要是用来消除静差，所谓静差，就是系统稳定后输出值和设定值之间的差值。积分环节实际上就是偏差累计的过程，把累计的误差加到原有系统上，以抵消系统造成的静差。而微分环节则反应了偏差信号的变化规律，或者说是变化趋势。根据偏差信号的变化趋势来进行超前调节，从而增加了系统的快速性。

6.6 The Parameter Tuning of the PID Controller

6.6 PID 控制器的参数整定

6.6.1 The Physical Meaning of the PID Parameters

6.6.1 PID 参数的物理意义

(1) The Understanding of the Proportional Action

(1) 对比例作用的理解

The proportional, integral and derivative parts of the controller output have clear physical meanings. Understanding their physical meaning helps to adjust the parameters of the controller. The principle of PID control can be understood by manual temperature control of a furnace.

控制器输出量中的比例、积分和微分部分都有明确的物理意义。了解它们的物理意义，有助于调整控制器的参数。PID 的控制原理可以用炉温的人工控制来理解。

Manual control is actually a closed-loop control. The operator reads the measured value of the temperature of the furnace detected by the digital instrument with eyes, and compares it with the set value of the temperature of the furnace to obtain the error value. Operate the potentiometer by hand to adjust the heating current to keep the temperature of the furnace near the set value. Experienced operators can obtain a good result by manual operation.

人工控制实际上也是一种闭环控制。操作人员用眼睛读取数字式仪表检测到的炉温的测量值，并与炉温的设定值比较，得到温度的误差值。用手操作电位器，调节加热电流，使炉温保持在设定值附近。有经验的操作人员手动操作后，可以得到很好的控制效果。

The operator knows the position of the potentiometer when the temperature of the furnace is stable at the set value (call it position L), and adjust the angle of the potentiometer according to the temperature error at that time. When the temperature of the furnace is less than the set value, the error is positive. Based on the position L, increase the angle of the potentiometer clock wise to increase the heating current. When the temperature of the furnace is greater than the set value, the error is negative. Based on the position L, decrease the angle of the potentiometer anticlockwise to reduce the heating current. Make the difference between the adjusted rotation angle of the potentiometer and the position L proportional to the error. The greater the absolute value of the error, the greater the adjusted angle. The above control strategy is the proportional control, i. e., the proportional part of the output of the PID controller is proportional to the error.

操作人员知道使炉温稳定在设定值时的电位器位置（将它称为位置 L），并根据当时的温度误差值调整电位器的转角。炉温小于设定值时，误差为正，在位置 L 的基础上顺时针增大电位器的转角，以增大加热的电流；炉温大于设定值时，误差为负，在位置 L 的基础上反时针减小电位器的转角，以减小加热的电流。令调节后的电位器转角与位置 L 的差值与误差成正比，误差绝对值越大，调节的角度越大。上述控制策略就是比例控制，即 PID 控制器输出中的比例部分与误差成正比。

There are various delays in the closed-loop. For example, there is a large delay between adjusting the angle of the potentiometer, and the temperature rising to the steady-state value corresponding to the new angle. There are delays in the thermal inertia of the heating furnace, the detection of temperature, the conversion from analog to digital, and the periodic calculation of PID. Due to the existence of the delay factor, the adjustment effect can not be seen immediately after adjusting the potentiometer angle.

闭环中存在着各种各样的延迟作用。例如调节电位器转角与温度上升到新的转角对应的稳态值之间有较大的延迟。加热炉的热惯性、温度的检测、模拟量转换为数字量和 PID 的周期性计算都有延迟。由于延迟因素的存在，调节电位器转角后不能马上看到调节的效果。

If the gain is too small, i. e., the difference between the adjusted rotation angle of the potentiometer and the position L is too small, the adjustment force will not be enough, which will cause slow temperature change and long adjustment time. If the gain is too large, i. e., the difference between the adjusted rotation angle of the potentiometer and the position L is too large, the adjustment force will be too strong, resulting in over-adjustment, which may cause the temperature to rise and fall, oscillate back and forth, and overshoot too much.

如果增益太小，即调节后电位器转角与位置 L 的差值太小，调节的力度不够，使温度的变化缓慢，调节时间过长。如果增益过大，即调节后电位器转角与位置 L 的差值过大，调节力度太强，造成调节过头，可能使温度忽高忽低，来回振荡，超调量过大。

If the closed-loop system has no integral effect, theoretical analysis shows that pure proportional control has steady-state error, which is inversely proportional to gain. When the system gain is small, the amount of overshoot and the number of oscillations are small, or there is no overshoot, but the steady-state error is large. When the gain is increased several times, the rising speed of the controlled variable is accelerated and the steady-state error is reduced. However, the overshoot increases, the oscillation times increase, the adjustment time is prolonged, and the dynamic performance becomes worse. The closed-loop system will become unstable. Therefore, it is difficult to take both dynamic and static performance into account with pure proportional control.

如果闭环系统没有积分作用，由理论分析可知，单纯的比例控制有稳态误差，稳态误差与增益成反比。当系统增益小时，超调量和振荡次数少，或者没有超调，但是稳态误差大。当增益增大几倍后，启动时被控量的上升速度加快，稳态误差减小；但是超调量增大，振荡次数增加，调节时间加长，动态性能变坏，甚至会使闭环系统不稳定。因此单纯的比例控制很难兼顾动态性能和静态性能。

(2) Understanding of the Integral Action

(2) 对积分作用的理解

Each integral operation of PID control in the computer is to add a tiny part proportional to the current error value on the basis of the original integral value (accumulated value of rectangular area). When the error is positive, the integral term increases; while the error is negative, the integral term decreases.

PID 控制在计算机中的每一次积分运算都是在原来的积分值（矩形面积的累加值）的基础上，增加一个与当前的误差值成正比的微小部分。误差为正时，积分项增大；误差为负时积分项减小。

In the above-mentioned temperature control system, integral control fine-tunes the angle of the potentiometer in each sampling period according to the current error value. As long as the error is not zero, the output

在上述的温度控制系统中，积分控制根据当时的误差值，每个采样周期都要微调电位器的角度。只要误差不为零，控制器的输出

of the controller will continue to change due to the integral action. The "general direction" of this fine-tune of integral control is correct. As long as the error is not zero, the integral term will change in the direction of reducing the error. When the error is small, the effects of the proportional part and the derivative part are almost negligible, but the integral term is still constantly changing. With the force of " dropping through the stone", the error will approach zero.

就会因为积分作用而不断变化。积分这种微调的“大方向”是正确的，只要误差不为零，积分项就会向减小误差的方向变化。在误差很小的时候，比例部分和微分部分的作用几乎可以忽略不计，但是积分项仍然不断变化，用“水滴石穿”的力量，使误差趋近于零。

When the system is in a stable state, the error is always zero, and the proportional part and the derivative part are both zero. At the same time, the integral part is no longer changed and just equal to the required output value of the controller in the steady state, corresponding to the position L of the potentiometer angle in the above temperature control system. The function of the integral part is to eliminate the steady-state error, improve the control accuracy, and make the controlled quantity eventually be equal to the set value. Therefore the integral part is generally required.

系统处于稳定状态时，误差恒为零，比例部分和微分部分均为零。同时，积分部分不再变化，并且刚好等于稳态时需要的控制器的输出值，对应于上述温度控制系统中电位器转角的位置L。积分部分的作用是消除稳态误差，提高控制精度，使被控量最终等于设定值。因此积分部分一般是必需的。

Although the integral control can eliminate the steady-state error. But if the parameters are not tuned well, the integral control also has a negative effect. If the integral action is too strong, it means that the angle value of the potentiometer for each trimming is too large. After accumulating into the integral term, it will make the dynamic performance of the system worse, increase the overshoot, and even make the system unstable. If the integral action is too weak, the error elimination speed is too slow.

积分虽然能消除稳态误差，但是如果参数整定得不好，积分也有负面作用。如果积分作用太强，相当于每次微调电位器的角度值过大，累积为积分项后，会使系统的动态性能变差，超调量增大，甚至使系统不稳定。积分作用太弱，则消除误差的速度太慢。

The integral term is different with the proportional term. It is accumulated by the current error value and the past error values. Therefore, the integral operation has serious hysteresis characteristics, which is detrimental to the stability of the system. If the integration time is not well set, it's negative effects are difficult to quickly correct through the integration effect itself.

积分项与比例项不同，它由当前误差值和过去的历次误差值累加而成。因此积分运算本身具有严重的滞后特性，对系统的稳定性不利。如果积分时间设置得不好，其负面作用很难通过积分作用本身迅速地修正。

The integral control with hysteresis characteristics is rarely used alone, it's generally used in conjunction with the proportional and derivative control to form the PI or PID controller. PI and PID controllers not only overcome the shortcomings of simple proportional adjustment with steady-state error, but also avoid the shortcomings of simple integral adjustment of slow response and poor dynamic performance. Therefore, they are widely used. If the controller has an integral function, the steady-state error of the step input can be eliminated, and then the gain can be reduced.

具有滞后特性的积分控制很少单独使用，它一般与比例控制和微分控制联合使用，组成 PI 或 PID 控制器。PI 和 PID 控制器既克服了单纯的比例调节有稳态误差的缺点，又避免了单纯的积分调节响应慢、动态性能不好的缺点，因此被广泛使用。如果控制器有积分作用，积分能消除阶跃输入的稳态误差，这时可以将增益调得小一些。

(3) Understanding of the derivative action

(3) 对微分作用的理解

The derivative component of the PID output is directly proportional to the change rate of error. The faster the error changes, the greater the absolute value of the derivative term. The sign of the derivative term reflects the direction of the error change. The derivative part of the output of the controller reflects the changing trend of the controlled quantity.

PID 输出的微分分量与误差的变化速率成正比，误差变化越快，微分项的绝对值越大。微分项的符号反映了误差变化的方向。控制器输出量的微分部分反映了被控量变化的趋势。

When the temperature rises too fast, but has not reached the set value, the experienced operator predicts that the temperature will exceed the set value according to the trend of the temperature change and the overshoot will occur. Therefore, the angle of potentiometer is adjusted to reduce the heating current in advance to reduce the overshoot.

有经验的操作人员在温度上升过快，但是尚未达到设定值时，会根据温度变化的趋势，预感到温度将会超过设定值，出现超调。于是调节电位器的转角，提前减小加热的电流，以减小超调量。

The fundamental reason for the oscillation and even instability of the closed-loop control system is that there is a large lag factor, and the leading effect of derivative control can offset the influence of the lag factor. Proper derivative control can reduce overshoot, shorten regulation time and increase system stability. If the control effect of PI controller is not ideal, the derivative action can be added to the controller to improve the dynamic characteristics of the closed-loop system.

闭环控制系统的振荡甚至不稳定的根本原因在于有较大的滞后因素，微分控制的超前作用可以抵消滞后因素的影响。适当的微分控制可以使超调量减小，调节时间缩短，增加系统的稳定性。如果 PI 控制器的控制效果不理想，可以考虑在控制器中增加微分作用，以改善闭环系统的动态特性。

(4) Determination of the sampling period

(4) 采样周期的确定

The PID control program is executed periodically. The smaller the sampling period T_s, the more the sampling value can reflect the change of the analog quantity. However, if T_s is too small, it will increase the computing workload of the CPU and cause system work delay or even step lose, so it's not suitable to get too small T_s.

PID控制程序是周期性执行的，采样周期 T_s 越小，采样值越能反映模拟量的变化情况。但是 T_s 太小会增加CPU的运算工作量而导致系统工作出现延迟甚至丢步，所以也不宜将 T_s 取得过小。

When determining the sampling period, it should be ensured that there are enough sampling points in the section where the controlled quantity changes rapidly (such as the rising stage of the start-up process). By connecting the process variables of each sampling point, the process variable curve of analog quantity should be basically reproduced to ensure that the important information in the analog quantity will not be lost due to the scarcity of sampling points.

确定采样周期时，应保证在被控量迅速变化的区段（例如启动过程的上升阶段）有足够多的采样点。将各采样点的过程变量连接起来，应能基本上复现模拟量的过程变量曲线，以保证不会因为采样点过稀而丢失模拟量中的重要信息。

The empirical data of sampling period in process control are given in Table 6-1. The data in the table are for reference only. Taking the temperature control as an example, the thermal inertia of a small incubator is much smaller than that of a heating furnace with tens of cubic meters. Obviously, their sampling periods should be greatly different. The actual sampling period needs to be determined after field debugging.

表6-1给出了过程控制中采样周期的经验数据，表中的数据仅供参考。以温度控制为例，一个很小的恒温箱的热惯性比几十立方米的加热炉的热惯性小得多，它们的采样周期显然也应该有很大的差别。实际的采样周期需要经过现场调试后确定。

Table 6-1 Empirical data of the sampling period

表6-1 采样周期的经验数据

Controlled amount 被控制量	Temperature 温度	Pressure 压力	Level 液位	Flow 流量	Element 成分
Sampling period/s 采样周期/s	15～25	3～10	6～8	1～5	15～20

6.6.2 The Method of PID Parameter Tuning

6.6.2 PID的参数整定方法

The parameter tuning of the PID controller is the core content of control system design. It's a process of de-

PID控制器的参数整定是控制系统设计的核心内容。它是根据被控过

termining the proportional amplification factor, integral time and derivative time constant of the PID controller according to the characteristics of the controlled process.

程的特性确定 PID 控制器的比例放大倍数、积分时间常数和微分时间常数大小的过程。

There are many methods of the parameter tuning of the PID controller, which can be summarized into two categories. One of them is the theoretical calculation setting method which determines the controller parameters through theoretical calculations mainly based on the mathematical model of the system. The calculated data obtained by this method may not be directly usable and must be adjusted and modified through actual engineering. The second is the engineering tuning method, which mainly relies on engineering experience and is carried out directly in the experiment of the control system. This method is simple and easy to master, and is widely used in engineering practice. The engineering tuning methods mainly include critical proportion method, response curve method and attenuation method. However, the controller parameters obtained by any method need to be adjusted and perfected in practice.

PID 控制器参数整定的方法很多，概括起来有两大类：一是理论计算整定法，它主要依据系统的数学模型，经过理论计算确定控制器参数。这种方法所得到的计算数据未必可以直接用，还必须通过工程实际进行调整和修改。二是工程整定方法，它主要依赖工程经验，直接在控制系统的实验中进行，且方法简单、易于掌握，在工程实际中被广泛采用。工程整定方法主要有临界比例法、反应曲线法和衰减法。但是，无论采用哪一种方法，所得到的控制器参数都需要在实际运行中进行最后调整与完善。

The Ziegler-Nichols tuning formula is a practical empirical formula for the first-order system with time delay. In this case, the system can be set as follows

Ziegler-Nichols 整定公式是针对带有时延环节的一阶系统而提出的实用经验公式。此时，可将系统设定为如下形式

$$G(s)=\frac{Ke^{-T_1 s}}{1+T_2 s} \tag{6-15}$$

In practical control systems, a large number of systems can be similar to Ziegler Nichols model. The following introduces several commonly used PID parameter setting methods in engineering.

在实际控制系统中，大量的系统可用 Ziegler-Nichols 模型近似，以下介绍几种工程中常用的 PID 参数整定方法。

(1) Open-loop Curve Method

(1) 开环曲线法

The step signal is applied to the controlled object (open-loop system), and it's response signal is measured by the experimental method, as shown in Fig. 6-10.

给被控对象（开环系统）施加阶跃信号，通过实验方法，测出其响应信号，如图 6-10 所示，由图

According to the shape of the output signal in the figure, the system response delay time T_1, the system time constant T_2, and the magnification factor $K = Y/X$ are approximately determined. After these three parameters are determined, the PID controller model can be determined according to Table 6-2, i. e., to set the relevant parameters in Eq. (6-15).

中输出信号的形状近似确定系统响应延时时间 T_1，系统时间常数 T_2，放大倍数 $K=Y/X$。确定这三个参数后，可根据表 6-2 确定 PID 控制器模型，即设定式(6-15)中有关参数。

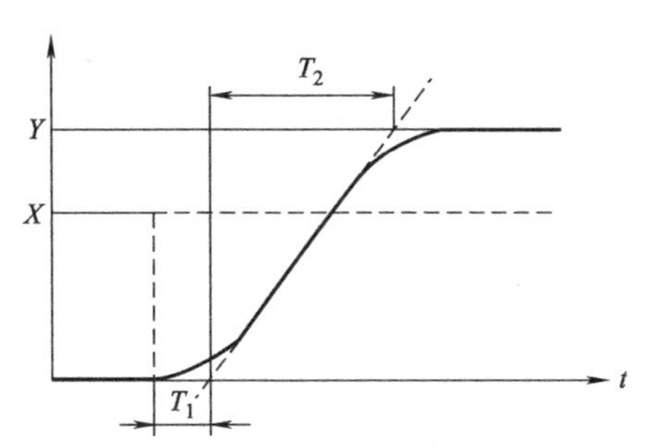

Fig. 6-10　Step response of the first-order delay system

图 6-10　一阶时延系统阶跃响应

Table 6-2　The parameter tuning table of the open-loop curve method

表 6-2　开环曲线参数整定表

Types 类型	K_P	T_I	T_D
P	$\frac{T_2}{KT_1}$	—	—
PI	$0.9\frac{T_2}{KT_1}$	$\frac{T_1}{0.3}$	—
PID	$1.2\frac{T_2}{KT_1}$	$2.2T_1$	$0.5T_1$

In some systems with fast response speed, such as small servo mechanism, T_1 may be very small, the effect of this method is relatively poor.

在某些响应速度快的系统中，比如小型伺服机构中，T_1 有可能很小，这种方法效果就比较差了。

【Example 6-1】 It is known that the open-loop transfer function of a control system is $G(s)=\frac{8e^{-180s}}{1+360s}$. Try to use the Ziegler-Nichols tuning formula to calculate the parameters of P, PI and PID controllers, and draw the unit step response curve of the system after setting.

【例 6-1】 已知某控制系统开环传递函数为 $G(s)=\frac{8e^{-180s}}{1+360s}$，试采用 Ziegler-Nichols 整定公式计算系统 P、PI、PID 控制器的参数，并绘制整定后系统的单位阶跃响应曲线。

Solution　PID parameter tuning is a process of repeated adjustment and testing, which can be greatly simplified by using Simulink model. According to the meaning of

解　PID 参数整定是一个反复调整测试的过程，使用 Simulink 模型能大大简化这一过程。根据题意，建

the question, the Simulink model as shown in Fig. 6-11 is established.

立如图 6-11 所示的 Simulink 模型。

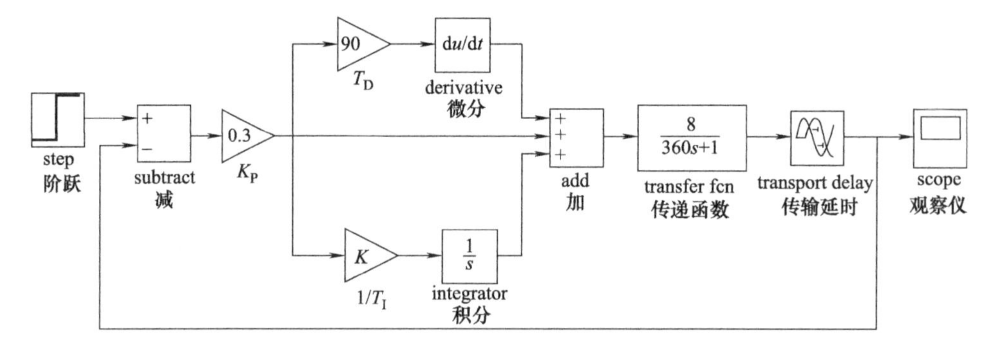

Fig. 6-11 Simulink model of the system

图 6-11 系统的 Simulink 模型

The first step of Ziegler-Nichols tuning is to obtain the unit step response of the open-loop system. In Simulink model, the feedback line, the output line of the differentiator and the output line of the integrator are disconnected. At the same time, set the value of "K_P" to 1, select the simulation time (note that if the system lag is large, the simulation time should be increased accordingly), and then run the simulation. After running, double-click "Scope" to get the result as shown in Fig. 6-12.

Ziegler-Nichols 整定的第一步是获取开环系统的单位阶跃的响应，在 Simulink 模型中，把反馈连线、微分器的输出连线、积分器的输出连线都断开，"K_P" 的值置为 1，选定仿真时间（注意，如果系统滞后比较大，则应相应加大仿真时间），仿真运行，运行完毕后，双击 "Scope"，得到如图 6-12 所示的结果。

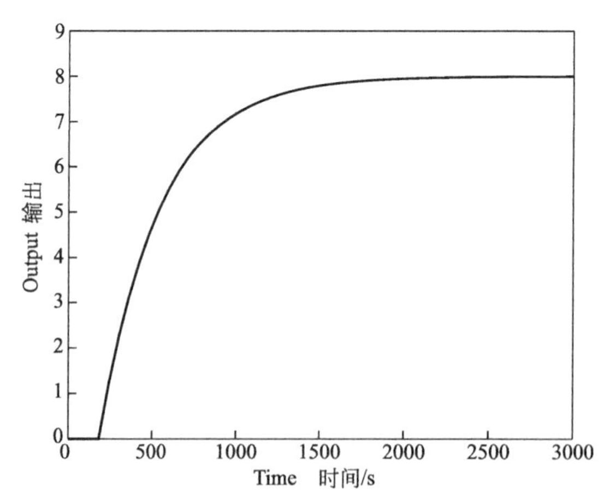

Fig. 6-12 Simulation results with $K_P=1$

图 6-12 $K_P=1$ 时的仿真结果

According to the parameter seeking method of the open-loop response curve, the delay time T_1, the amplification factor K and the time constant T_2 can be substantially achieved as follows

按照开环响应曲线的参数求法，可以得到系统延迟时间 T_1，放大系数 K 和时间常数 T_2 为

$$T_1=180, \ T_2=360, \ K=8$$

If it is not easy to see these three parameters from the output of the oscilloscope, the system output can be imported into the Matlab workspace, and then write the corresponding m file to get these three parameters.

如果从示波器的输出不易看出这 3 个参数，则可以将系统输出导入到 Matlab 的工作空间中，然后编写相应的 m 文件求取这 3 个参数。

According to table 6-2, the proportional amplification factor K_P is 0.25 when P control is tuning. Set the value of "K_P" to 0.25 and run the simulation. After running, double-click "Scope" to get the result as shown in Fig. 6-13, which is the unit step response of the system under P control.

根据表 6-2，可知 P 控制整定时，比例放大系数 $K_P=0.25$，将"K_P"的值置为 0.25，仿真运行，运行完毕后，双击"Scope"得到图 6-13 所示的结果，它是 P 控制时系统的单位阶跃响应。

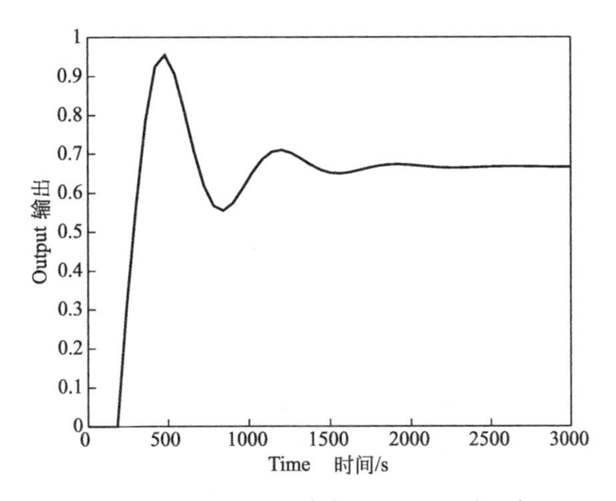

Fig. 6-13 Unit step response of the system under the P control

图 6-13 P 控制时系统的单位阶跃响应

It can be seen that the proportional amplification factor K_P is 0.225 and the integral time constant T_I is 600 when PI control is tuning. Set the value of "K_P" to 0.225 and the value of "$1/T_I$" to 1/600. Connect the output line of the integrator and run the simulation. After running, double-click "Scope" to get the result as shown in Fig. 6-14, which is the unit step response of the system under PI control.

当 PI 控制整定时，比例放大系数 $K_P=0.225$，积分时间常数 $T_I=600$，将"K_P"的值置为 0.225，"$1/T_I$"的值置为 1/600，将积分器的输出连线连上，仿真运行，运行完毕后双击"Scope"得到图 6-14 所示的结果，它是 PI 控制时系统的单位阶跃响应。

It can be seen that the proportional amplification coefficient $K_P=0.3$, the integral time constant $T_I=396$, and the differential time constant $T_D=90$. Set the value of "K_P" to 0.3, "$1/T_I$" to 1/396, and "T_D"

当 PID 控制整定时，比例放大系数 $K_P=0.3$，积分时间常数 $T_I=396$，微分时间常数 $T_D=90$，将"K_P"的值置为 0.3，"$1/T_I$"的

to 90. Connect the output line of the differentiator and run the simulation. After running, double-click "Scope" to get the result as shown in Fig. 6-15, which is the unit step response of the system during PID control.

值置为 1/396，"T_D" 值置为 90，将微分器的输出连线连上，仿真运行，运行完毕后，双击 "Scope" 得到如图 6-15 所示的结果，它是 PID 控制时系统的单位阶跃响应。

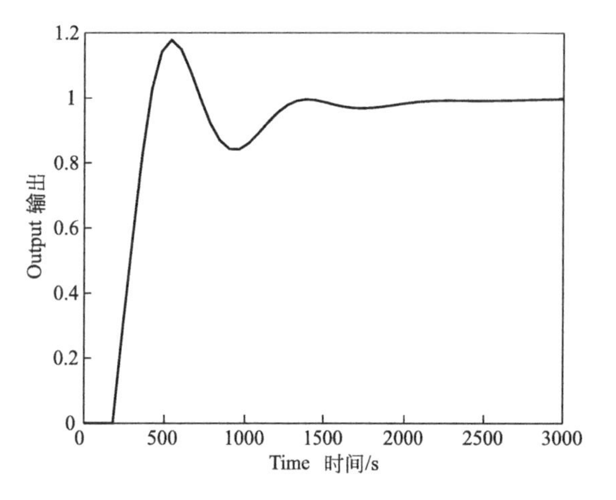

Fig. 6-14 Unit step response of the system under the PI control

图 6-14 PI 控制时系统的单位阶跃响应

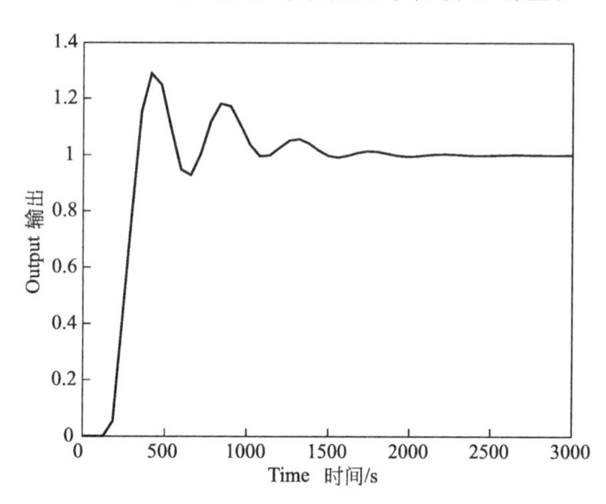

Fig. 6-15 Unit step response of the system under the PID control

图 6-15 PID 控制时系统的单位阶跃响应

Compared with Fig. 6-13 ~ Fig. 6-15, it can be seen that the response speed of the P control and PI control is basically same. Because the proportional coefficients of the two controllers are different, the stable output value of the system is different. The overshoot of PI control is smaller than that of P control, and the response speed of PID control is faster than that of P control and PI control while it's overshoot is larger.

由图 6-13～图 6-15 对比可知，P 控制和 PI 控制两者的响应速度基本相同，因为两种控制的比例系数不同，因此系统稳定的输出值不同。PI 控制的超调量比 P 控制的要小，PID 控制比 P 控制和 PI 控制的响应速度要快，但是超调量要大些。

(2) Closed-loop Limit Period Method

(2) 闭环极限周期法

The method is to make the system as a closed-loop system with only proportional control, that is, set $T_I=\infty$, $T_D=0$, and apply a step signal to the system. Initially set K_P to be small, and then gradually increase it until the closed-loop system produces a constant amplitude oscillation, which is shown in Fig. 6-16. Record the magnification at this time as K_u and the critical oscillation period as T_u, and then set the PID parameters according to Table 6-3.

该方法是令系统为只有比例控制的闭环系统，即令 $T_I=\infty$，$T_D=0$，给系统施加一个阶跃信号，初始时令 K_P 较小，然后逐渐增大，直至闭环系统产生等幅振荡，如图 6-16 所示。记录此时的放大倍数 K_u 和临界振荡周期 T_u，然后根据表 6-3 整定 PID 参数。

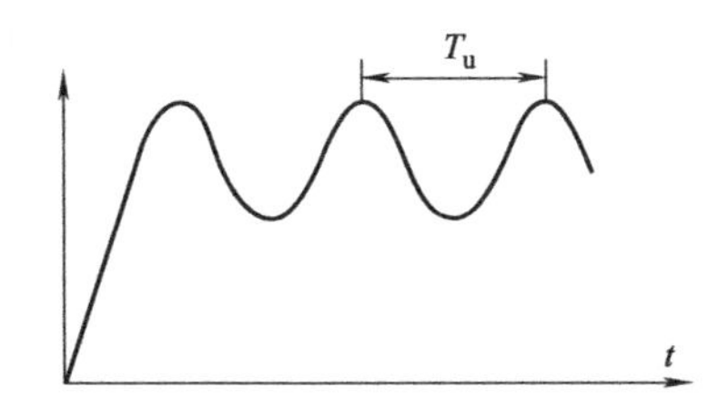

Fig. 6-16 Constant oscillation response

图 6-16 等幅振荡响应

Table 6-3 Parameter tuning table of the closed-loop limit cycle method

表 6-3 闭环极限周期法参数整定表

Type 类型	K_P	T_I	T_D
P	$0.5K_u$	—	—
PI	$0.45K_u$	$\frac{T_u}{1.2}$	—
PID	$0.6K_u$	$\frac{T_u}{2}$	$\frac{T_u}{8}$

Sometimes constant amplitude oscillation is not allowed in engineering practice, or constant amplitude oscillation within the normal operating range cannot be generated, so this method cannot be used.

工程实际中有时不允许出现等幅振荡，或在正常操作范围内无法产生等幅振荡，此时无法使用该方法。

【Example 6-2】 It is known that the open-loop transfer function of a control system is $G(s) = \frac{1}{s(s+1)(s+5)}$. Try to use the tuning formula of the closed-loop limit period to calculate the parameters of the P, PI and PID controllers, and draw the unit step response curve of the system after tuning.

【例 6-2】 已知某控制系统开环传递函数为 $G(s) = \frac{1}{s(s+1)(s+5)}$，试采用闭环极限周期整定公式计算系统 P、PI、PID 控制器的参数，并绘制整定后系统的单位阶跃响应曲线。

Solution According to the meaning of the question, the Simulink model shown in Fig. 6-17 can be established.

解 根据题意，建立如图 6-17 所示的 Simulink 模型。

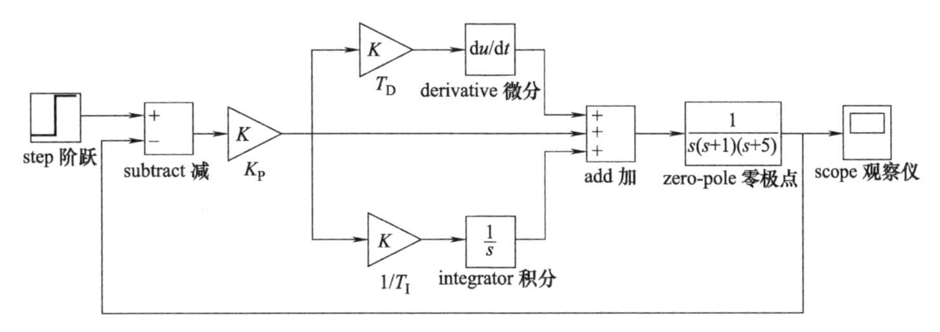

Fig. 6-17 Simulink model of example 6-2

图 6-17 例 6-2 Simulink 模型

The first step of the tuning method of the closed-loop limit period is to obtain the constant amplitude oscillation curve of the system. In Simulink, disconnect the output connections of the differentiator, and the the integrator, and test the value of "K_P" from large to small (small to large). After each simulation, observe the output of the oscilloscope until the constant amplitude oscillation curve is output. In this example, constant amplitude oscillation occurs when K_P = 30. At this time, the constant amplitude oscillation curve of T_u=2.9 is shown in Fig. 6-18.

闭环极限周期整定法的第一步是获取系统的等幅振荡曲线。在Simulink中，把微分器的输出连线、积分器的输出连线都断开，令"K_P"的值从大到小进行试验(小到大)，每次仿真结束后，观察示波器的输出，直到输出等幅振荡曲线为止。本例中当K_P=30时出现等幅振荡，此时T_u=2.9的等幅振荡曲线如图6-18所示。

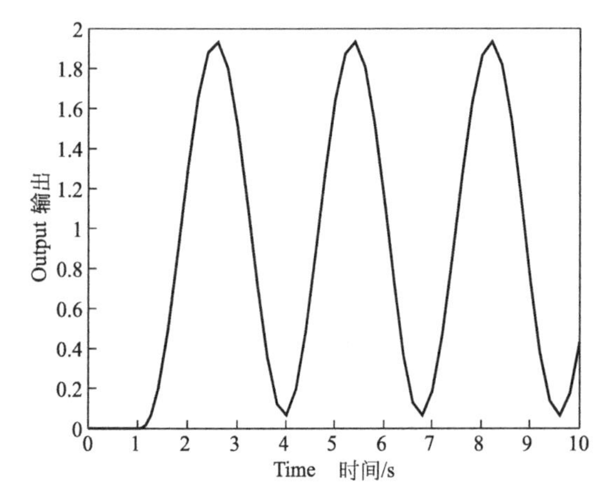

Fig. 6-18 Constant oscillation curve of the system

图6-18 系统的等幅振荡曲线

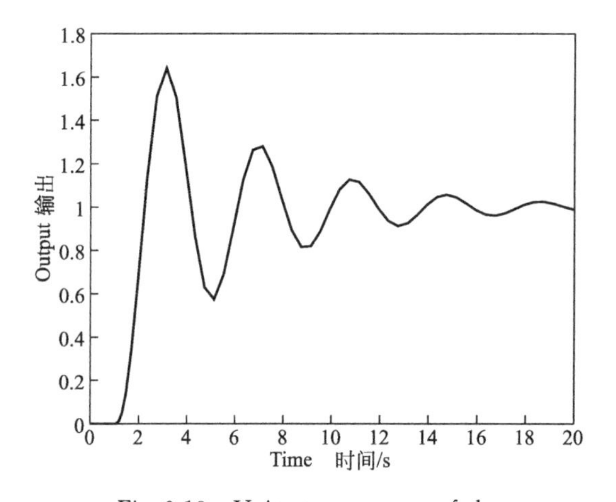

Fig. 6-19 Unit step response of the system under the P control

图6-19 P控制时系统的单位阶跃响应

According to Table 6-3, it can be seen that the proportional amplification factor K_P=15 when P control is tuning. Set the value of "K_P" to 15 and run the simulation. After running, double-click "Scope" to get the result shown in Fig. 6-19, which is the unit step response of the system under the P control.

根据表6-3，可知P控制整定时，比例放大系数K_P=15，将"K_P"的值置为15，仿真运行，运行完毕后，双击"Scope"，得到如图6-19所示的结果，它是P控制时系统的单位阶跃响应。

It can be seen that the proportional amplification factor K_P=13.5 and the integral time constant T_I=2.42 when PI control is tuning. Set the value of "K_P" to 13.5 and the value of "$1/T_I$" to 1/2.42. Then connect the output wire of the integrator and run the simulation. After running, double-click "Scope" to get the result shown in Fig. 6-20, which is the unit step

当PI控制整定时，比例放大系数K_P=13.5，积分时间常数T_I=2.42。将"K_P"的值置为13.5，"$1/T_I$"的值置为1/2.42，并将积分器的输出连线连上，仿真运行，运行完毕后双击"Scope"，得到如图6-20所示的结果，它是PI控制

response of the system under the PI control.

时系统的单位阶跃响应。

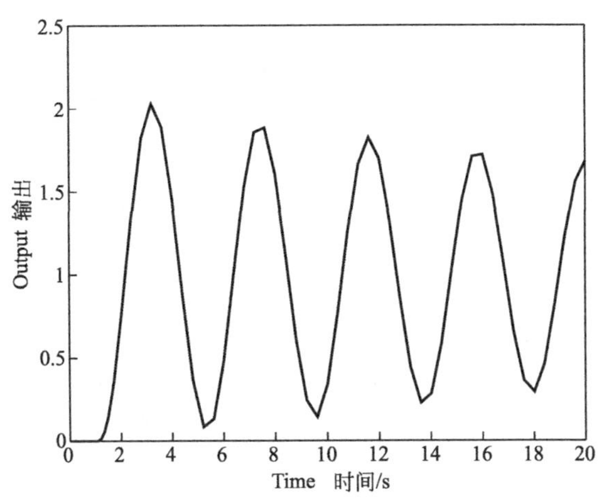

Fig. 6-20 Unit step response of the system under the PI control

图 6-20 PI 控制时系统的单位阶跃响应

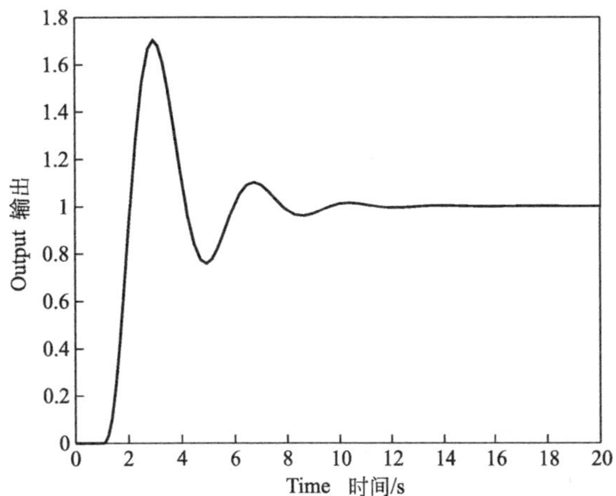

Fig. 6-21 Unit step response of the system under the PID control

图 6-21 PID 控制时系统的单位阶跃响应

It can be seen that the proportional amplification factor $K_P = 18$, the integral time constant $T_I = 1.45$, and the derivative time constant $T_D = 0.3625$ when PID control is tuning. Set the value of "K_P" to 18, the value of "$1/T_I$" to 1/1.45, and the value of "T_D" to 0.3625. Then, connect the output wire of the differentiator and run the simulation. After running, double-click "Scope" to get the result shown in Fig. 6-21, which is the unit step response of the system under the PID control.

当 PID 控制整定时，比例放大系数 $K_P = 18$，积分时间常数 $T_I = 1.45$，微分时间常数 $T_D = 0.3625$。将"K_P"的值置为 18，"$1/T_I$"的值置为 1/1.45，"T_D"值置为 0.3625，并将微分器的输出连线连上，仿真运行，运行完毕后，双击"Scope"，得到如图 6-21 所示的结果，它是 PID 控制时系统的单位阶跃响应。

Compared with Fig. 6-19 ~ Fig. 6-21, it can be seen that the response speed of the P control and PI control is basically same, but the stable output value of the system is different. The overshoot of PI control is bigger than that of P control, and the response speed of PID control is faster than that of P control and PI control while its overshoot is larger.

由图 6-19～图 6-21 对比可知，P 控制和 PI 控制的阶跃响应速度基本相同，但系统稳定的输出值不同。PI 控制的超调量比 P 控制的要大，PID 控制比 P 控制和 PI 控制的响应速度要快，但是超调量要大些。

It is worth noting that the engineering tuning method is based on empirical formula, which is not applicable in any case. Therefore, the PID parameters tuning according to the empirical formula is not the best and needs to be adjusted. In this example, the parameter

值得注意的是，由于工程整定方法依据的是经验公式，不是在任何情况下都适用的，因此，按照经验公式整定的 PID 参数并不是最好的，需要进行一些调整。本例

tuning of the PI controller according to Table 6-3 is not very good, as can be seen from the figure. Adjust the proportional amplification factor K_P to 13.5 and the integral time constant T_I to 12.5, and then run the simulation. After running, double-click "Scope" to get the result as shown in Fig. 6-22.

中，按照表 6-3 整定的 PI 控制器的参数就不是非常好，这从图中可以看出。将比例放大系数调整为 $K_P=13.5$，积分时间常数 T_I 调整为 12.5，仿真运行，运行完毕后，双击"Scope"得到如图 6-22 所示的结果。

It can be seen from Fig. 6-20 and Fig. 6-22 that the overshoot of the system is reduced and the adjustment time is also reduced after adjusting the parameters of the PI. Of course, there are many ways to adjust parameters, you can adjust the parameters of P or I or both of them at the same time.

对比图 6-20 和图 6-22 可知，调整 PI 参数后系统的超调量减小了，调整时间也减小了。当然，调整参数的方法有多种，既可以调整 P 的参数，也可以调整 I 的参数，也可以同时调整这两者的参数。

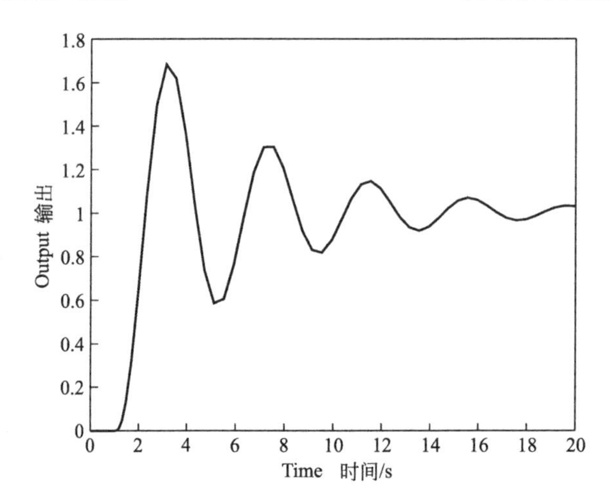

Fig. 6-22 Unit step response after adjusting the parameters of PI

图 6-22 系统调整 PI 参数后的单位阶跃响应

(3) Empirical Method

(3) 经验法

The empirical method is also called field test method, that is, determine the parameter values K_P and T_I of a regulator firstly, then apply a disturbance to the control system by changing the given value, and observe and judge the shape of the control curve on the spot. If the curve is not ideal, K_P or T_I can be changed to control the process curve. Test repeatedly until the control system meets the quality requirements of dynamic process, in which case K_P and T_I are the best values.

经验法又叫现场试验法，即先确定一个调节器的参数值 K_P 和 T_I，通过改变给定值对控制系统施加扰动，现场观察判断控制曲线形状。若曲线不够理想，则改变 K_P 或 T_I，再观察控制过程曲线。反复测试，直到控制系统符合动态过程品质要求为止，这时的 K_P 和 T_I 就是最佳值。

If the controller is a PID controller, then the differential action should be added on the basis of the tuned K_P and T_I. The differential effect has the ability to resist the deviation change. Therefore, after determining the value of T_D, you can reduce the points of K_P and T_I that have been tuned for field trials until K_P, T_I and T_D achieve the best values. Obviously, the parameters tuned by the empirical method are accurate, but it takes more time. In order to shorten the tuning time, we should note the following:

如果控制器是 PID 控制器，那么要在整定好的 K_P 和 T_I 的基础上加微分作用。由于微分作用有抵制偏差变化的能力，所以确定 T_D 值后，可把整定好的 K_P 和 T_I 减小一点，再进行现场试凑，直到 K_P、T_I 和 T_D 取得最佳值为止。显然用经验法整定的参数是准确的，但花时间较多。为缩短整定时间，应注意以下几点：

• Determine the initial parameter value according to the characteristics of the controlled object. The determination of K_P, T_I and T_D can refer to the parameter values of similar control systems in actual operation or the parameter values given in Table 6-4, so as to make the determined initial parameters as close to the ideal values as possible. This can greatly reduce the number of field trials.

• 根据控制对象特性确定初始参数。K_P、T_I 和 T_D 可参照实际运行中的同类控制系统的参数值，或参考表 6-4 中的参数值，使确定的初始参数尽量接近理想值。这样可大大减少现场凑试的次数。

• In the process of trial, if it is found that the controlled variable changes slowly and can not reach the stable value as soon as possible, which is caused by excessive K_P or too long T_I. But there is a difference between the two. When K_P is excessive, the curve floats greatly and changes irregularly. However, if T_I is too long, the curve has oscillation component, and it approaches the given value very slowly. K_P or T_I can be changed according to curved shape.

• 在凑试过程中，被控量变化缓慢，不能尽快达到稳定值，可能是由于 K_P 过大或 T_I 过长。但两者是有区别的：K_P 过大，曲线浮动较大，变化不规则；而 T_I 过长，曲线带有振荡分量，且非常缓慢地接近给定值。可根据曲线形状来改变 K_P 或 T_I。

• If K_P is too small, T_I is too short, and T_D is too long, the oscillation will slow down or even not attenuate. If K_P is too small, the oscillation period is shorter. If T_I is too short, the oscillation period will be longer. If T_D is too long, the oscillation period is the shortest.

• K_P 过小，T_I 过短，T_D 太长都会导致振荡衰减得慢，甚至不衰减，其区别是 K_P 过小，振荡周期较短；T_I 过短，振荡周期较长；T_D 太长振荡周期最短。

• If the constant amplitude oscillation occurs in the tuning process and the phenomenon can not be eliminated by changing the parameters of the controller, it may be due to the incrorrect adjustment of the valve

• 如果在整定过程中出现等幅振荡，并且通过改变控制器参数也不能消除这一现象时，可能是阀门定位器调校不准，调节阀传动部

positioner, the clearance of the driving part of the control valve (or the size of the control valve is too large) or the interference of the control object by the constant amplitude fluctuation, etc. At this time, it is not only necessary to pay attention to the adjustment of the controller parameters, but to check and adjust other instruments and links.

分有间隙（或调节阀尺寸过大）或控制对象受到等幅波动的干扰等。这时就不能只注意控制器参数的整定，而是要检查与调校其他仪表和环节。

Table 6-4 The reference table of empirical data of PID

表 6-4 PID 参数经验数据参照表

Parameter 参数	Temperature 温度	Pressure 压力	Level 液位	Flow 流量
$K_P/\%$	20～60	30～70	20～80	40～100
T_I/s	180～600	24～180	60～300	6～60
T_D/s	3～180	—	—	—

The formula for adjusting PID parameters via the empirical method is as follows:

经验法的 PID 参数调节口诀如下：

Find the optimal parameters when parameters tuning, and check them in order from small to large;
First adjust the proportional control, then the integral control, and finally the derivative control;
When the curve oscillates frequently, the proportional dial should be enlarged;
When the curve floats around the big bay, the proportional dial should be moved down;
When the curve deviates and the recovery is slow, the integration time should decrease;
If the fluctuation period of the curve is long, the integration time will be longer;
If the oscillation frequency of the curve is fast, the derivative should be lowered first;
When the dynamic difference is large and the fluctuation is slow, the derivative time should be lengthened;
The ideal curve has two waves, high in the front and low in the back, and their amplitude ratio is 4 to 1;
Look first, then adjust, and analyze again, the quality of the adjustment will not be low.

参数整定找最佳，从小到大顺序查；
先是比例后积分，最后再把微分加；
曲线振荡很频繁，比例度盘要放大；
曲线漂浮绕大弯，比例度盘往小扳；
曲线偏离回复慢，积分时间往下降；
曲线波动周期长，积分时间再加长；
曲线振荡频率快，先把微分降下来；
动差大来波动慢，微分时间应加长；
理想曲线两个波，前高后低 4 比 1；
一看二调多分析，调节质量不会低。

(4) Attenuation Curve Method

(4) 衰减曲线法

We know that when the pure proportional control is used in the control system, the transition process as shown in Fig. 6-23 will appear when the proportional gain is gradually reduced.

当控制系统使用纯比例控制时，在比例增益逐步减少的过程中，就会出现图 6-23 所示的过渡过程。

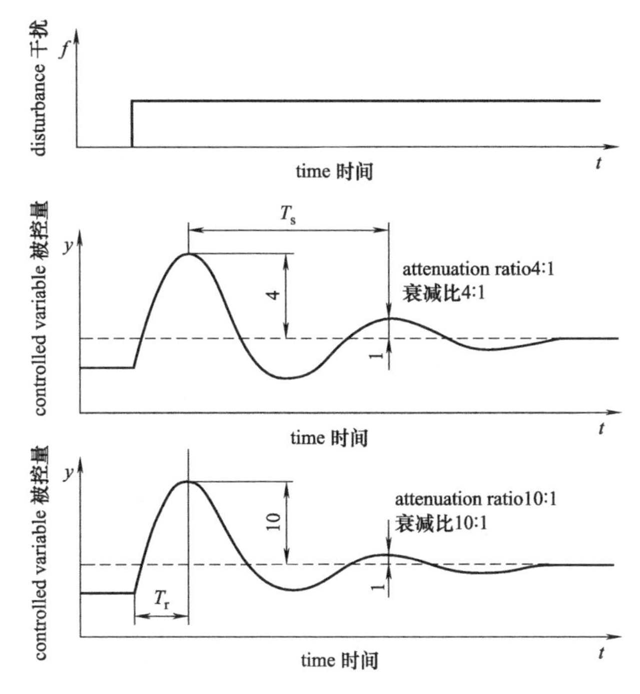

Fig. 6-23 The attenuation curve of transition process with step change of given value

图 6-23 给定值阶跃变化下的过渡过程衰减曲线

At this time, the amplitude ration of two adjacent peaks in the control process is $n:1$, which is called attenuation ratio δ_s, and the distance between the two wave peaks is called the attenuation period T_s. The attenuation curve method is to obtain the attenuation ratio δ_s and the attenuation period T_s in a purely proportional control system. And based on these data, the parameters of the controller, include T_I and T_D are calculated. Detail tuning process is presented as follows:

这时控制过程相邻两个波峰的幅值比为 $n:1$，称为衰减比 δ_s；两个波峰之间的距离，称为衰减周期 T_s。衰减曲线法，就是在纯比例作用的控制系统中，求得衰减比 δ_s 和衰减周期 T_s，并依据这两个数据来计算出控制器的参数 T_I 和 T_D。具体整定过程如下：

Attenuation tuning has many advantages, which is not only safe but also fast;

Pure P reduces the attenuation ration until finds the attenuation ration 4 : 1;

According to the formula to calculate, and then increase the PID parameters in order;

衰减整定好处多，操作安全又迅速；

纯 P 降低衰减比，找到衰减比 4 : 1；

按照公式来计算，PID 序加参数；

Watch the operation and adjust carefully until the best value is found.

观看运行细调整，直到找出最佳值。

• Firstly, the integral time is set to the maximum and the differential time is set to zero, so that the control system can operate and the proportional degree is set to a larger appropriate value. "Pure P reduces the attenuation ratio" means that the control system operates in pure proportion mode. Then slowly reduce the proportional degree, and observe the output of the controller and the fluctuation of the control process until the attenuation process of 4∶1 is found. This process is called "find attenuation ration 4∶1".

•先把积分时间放至最大，微分时间放至零，使控制系统运行，比例度放至较大的适当值。"纯P降低衰减比"，就是使控制系统按纯比例作用的方式投入运行。然后慢慢地减少比例度，观察控制器的输出及控制过程的波动情况，直到找出4∶1的衰减过程为止。这一过程就是"找到衰减比4∶1"。

• For some control objects, strong oscillation may occur when the attenuation ratio is 4∶1. At this time, the attenuation ratio of 10∶1 can be used. But it is very difficult to measure the attenuation period. At this time, the rise time T_r of the first wave peak can be measured, and the operation procedure is the same as above.

•对有些控制对象，用4∶1的衰减比感觉振荡过强时，可采用10∶1的衰减比。但这时要测量衰减周期是很困难的，可采取测量第一个波峰的上升时间 T_r，其操作步骤同上。

• Calculating each parameter value according to the attenuation ratio δ_s, the attenuation periods T_s, T_r and Table 6-5.

•根据衰减比 δ_s 和衰减周期 T_s、T_r，按表6-5进行计算，求出各参数值。

Table 6-5 The reference table of empirical data of PID

表6-5 PID参数经验数据参照表

Requirements of control quality 控制品质要求	Control law 控制律	$\delta/\%$	T_I/min	T_D/min
Attenuation ratio: 4∶1 衰减比为4∶1	P	δ_s	—	—
	PI	$1.2\delta_s$	$0.5T_I$	—
	PID	$0.8\delta_s$	$0.3T_I$	$0.1T_D$
Attenuation ratio: 10∶1 衰减比为10∶1	P	δ_s'	—	—
	PI	$1.2\delta_s'$	$2T_r$	—
	PID	$0.8\delta_s'$	$0.3T_r$	$0.1T_r$

• The proportional gain is placed on a value larger than the calculated value. Then, add the integration time T_I. And then slowly add the differential time T_D. When operating, you must follow the "increase

•将比例增益放在一个比计算值大的数值上，然后加上积分时间 T_I，再慢慢加上微分时间 T_D。操作时一定要按"PID序加参数"，即

PID parameters in order", that is, P first, I second, and D last. Do not break this sequence.

先P次I最后D，不要破坏了这个次序。

• Reduce the proportional gain to the calculated value, observe the curve, and then adjust the parameters appropriately. That is, watch the operation and adjust carefully until the best value is found.

•把比例增益降到计算值上，通过观察曲线，再适当调整各参数。即“观看运行细调整，直到找出最佳值”。

Many instrument workers have such experience: in the field of engineering parameter tuning of the controller, if only according to the attenuation ratio, then there can be many proportional gain and integration time constants that meet the attenuation ratio. But these values are not arbitrarily combined, but in pairs. A certain degree of proportional gain must form a pair with a certain integral time to satisfy the condition of attenuation ratio. Change one of them, and the other will also change accordingly. Because they appear in pairs, there is a problem of "matching" of controller parameters. In practical application, only by adding additional conditions can a pair of suitable values be selected from many pairs of values. This pair of suitable values is the real "best tuning value".

很多仪表工都有这样的体会，在现场的控制器工程参数整定中，如果只按衰减比进行整定，那么有很多比例增益和积分时间常数同时满足衰减比，但是这些数值并不是任意地组合，而是成对的。一定的比例增益必须与一定的积分时间组成一对，才能满足衰减比的条件。改变其中之一，另一个也要随之改变。因为参数成对出现，所以才有控制器参数的“匹配”问题。而在实际应用中，只有增加个附加条件，才能从多对数值中选出一对适合的值。这一对适合的值才是真正的“最佳整定值”。

(5) Extended Critical Proportional Gain Method

（5）扩充临界比例增益法

The extended critical proportional gain method is a parameter tuning method of the digital PID controller based on the critical proportional gain used in the analog PID controller. It is suitable for the controlled object with self balance and does not need the mathematical model of the controlled object. When the method is applied, the controllability must be determined first

扩充临界比例增益法是以模拟PID控制器中的临界比例增益为基础的一种数字PID控制器参数整定方法，它适用于具有自平衡性的被控对象，不需要被控对象的数学模型。应用扩充临界比例增益法时，首先要确定控制度

$$\underset{\text{控制度}}{\text{controllability}} = \frac{\left[\int_0^{\infty} e^2(t)\,\mathrm{d}t\right]_{\text{digit}}}{\left[\int_0^{\infty} e^2(t)\,\mathrm{d}t\right]_{\text{analog}}} \tag{6-16}$$

The controllability takes the error square integral as the evaluation function, which reflects the considera-

控制度以误差平方积分作为评价函数，反映了数字控制的控制效果

ble degree of the control effect of the digital control to that of the analog control. Due to the control delay caused by the digital control, the effect of the digital control is reduced when the same control law as that of the analog control is adopted. Moreover, the larger the sampling period T_s, the lower the control effect. However, the advantage of digital control is that it can choose control algorithm flexibly.

与模拟控制的控制效果的相当程度。由于数字控制造成的控制延时，使在采用与模拟控制相同的控制规律时，数字控制效果有所降低，而且采样周期 T_s 越大，控制效果降低越明显。但是，数字控制的优势在于可以灵活的选择控制算法。

Generally, when the controllability is 1.05, the control effect of the digital control is equivalent to that of the analog control; when the controllability is 2, the control effect of the digital control is twice as that of the analog control. In order to make the control effect of the digital PID controller as close as possible to the analog PD controller, the controllability should be close to 1.05. The steps of PID parameter tuning via the extended critical proportional gain method are as follows:

通常，控制度为 1.05 时，数字控制的控制效果与模拟控制相当；当控制度为 2 时，数字控制较模拟控制的控制效果差一倍。为使数字 PID 控制器的控制效果尽可能接近模拟 PD 控制器，应使控制度接近 1.05。用扩充临界比例增益法整定 PID 参数的步骤如下：

• Selecting a sufficiently short sampling period T_s. When there is a lag in the controlled process, the sampling period T_s is 1/10 or less of the time lag. In this case, the controller act as the pure proportional control.

• 选择一个足够短的采样周期 T_s。被控过程有滞后时，采样周期 T_s 取滞后时间的 1/10 以下，控制器作纯比例控制。

• Making the system work with the selected T_s. At this time, remove the integral and differential action and select the control as a pure proportional control, which constitutes a closed-loop operation mode. Under the step signal input, the K_P is gradually increased to make the control system appear in a critical oscillation state. Generally, the step response of the system continues 4 to 5 oscillations and the system is considered to have reached the critical oscillation state. Noting the K_P as the critical proportional coefficient K_r at this time, and the critical proportional degree is obtained as $\delta_r = 1/K_r$. The time from the first oscillation peak to the second peak is the oscillation period T_r.

• 用选定的 T_s 使系统工作。这时去掉积分作用和微分作用，将控制选择为纯比例控制，构成闭环运行。在阶跃信号输入下，逐渐加大 K_P，使控制系统出现临界振荡状态，一般系统的阶跃响持续 4～5 次振荡，就认为系统已到临界振荡状态。记此时的 K_P 为临界比例系数 K_r，得到临界比例度为 $\delta_r = 1/K_r$。从第一个振荡顶点到第二个顶点的时间为振荡周期 T_r。

• Selecting the controllability. The length of the sampling period T_s will affect the quality of the sampling data control system. With the same optimal setting, the control quality of the sampling data control system is lower than that of the continuous time control system. Therefore, the controllability is always greater than 1. And the greater the controllability, the worse the quality of the corresponding sampledata control system. The selection of the controllability should according to the requirements for the control quality of the designed system.

• 选择控制度。采样周期 T_s 的长短会影响采样数据控制系统的品质。在同样优化设置下，采样数据控制系统的控制品质要低于连续时间控制系统。因而控制度总是大于 1 的，而且控制度越大，相应的采样数据控制系统的品质越差。控制度的选择要从所设计的系统的控制品质要求出发。

• Detemining the parameters by looking up the table. According to the controllability and the Table 6-6, determine the values T, K_P, T_I and T_D.

• 查表确定参数。根据控制度，按表 6-6 选取 T、K_P、T_I 和 T_D 的值。

Table 6-6 Parameter table of extended critical proportional degree method

表 6-6 扩充临界比例度法整定参数表

Controllability 控制度	Control law 控制律	T/T_r	K_P/K_r	T_I/T_r	T_D/T_r
1.05	PI	0.03	0.53	0.88	—
	PID	0.14	0.63	0.49	0.14
1.20	PI	0.05	0.49	0.91	—
	PID	0.043	0.47	0.47	0.16
1.50	PI	0.14	0.42	0.99	—
	PID	0.09	0.34	0.43	0.20
2.00	PI	0.22	0.36	1.05	—
	PID	0.16	0.27	0.40	0.22
continuous time controller 连续时间控制器	PI	—	0.57	0.83	—
	PID	—	0.70	0.50	0.13

• Operation and correction. According to the obtained tuning parameters, run the system, observe the control effect, and then adjust the parameters appropriately until the satisfactory control effect is obtained.

• 运行与修正。按照求得的整定参数，投入系统运行，观察控制效果，再适当调整参数，直到获得满意的控制效果为止。

Chapter 7 Robot
第7章 机器人

Key words 重点词汇

industrial robot 工业机器人
robot arm 机械臂
six-axis robot 六轴机器人
robot specification 机器人规格
robot location 机器人位姿
robot program 机器人编程
sequential operation 顺序操作
conditional operation 循环操作
iterative operation 迭代操作
loop structure 循环结构
joint 关节

7.1 Introduction of Robot

7.1.1 Brief History of Robot

- 1921: The word robota was first used in a play for a slave performing compulsory tasks.
- 1962: The first generation robot, developed by Unimation, was pneumatic driven.
- 1975-1997: The second generation robots—programmable robots controlled by PLC that can operate in an orderly environment.
- 1990-present: The third generation robots, intelligent robots in unstructured environments (artificial intelligence).

The International Federation of Robotics has up to date information on the number and use of robots worldwide and also the history of the development of robots at related website.

7.1 机器人简介

7.1.1 机器人简史

- 1921年：戏剧中出现了robota，其作为奴隶执行必要任务。
- 1962年：第一代机器人，Unimation公司研制，气动驱动。
- 1975～1997：第二代机器人——可编程机器人，PLC控制，可以在一个有序的环境下操作。
- 1990～至今：第三代机器人，非结构化环境（人工智能）中的智能机器人。

国际机器人联合会拥有全球范围内机器人数量和用途的最新信息，以及有关机器人开发历史的信息，可以参考相关网站。

7.1.2 Definition of Robot

The definition of a robot is changing with develop ments in software capabilities. The definitions below are for an industrial robot.

- ISO: automatically controllable, reprogrammable, multipurpose manipulator programmable in three or more axes which may be either fixed in place or mobile for use in industrial automation applications.

- The Robotics Institute of America (RIA): reprogrammable multifunctional manipulator designed to move materials, parts, tools or specialized devices through variable programmed motions for the performance of a variety of tasks.

- British Robot Association (BRA): a reprogrammable device designed to both manipulate and transport parts, tools or specialized manufacturing implements through variable programmed motions for the performance of specific manufacturing tasks.

International Federation of Robotics (IFR) defines service robots as follows:

A service robot is a robot that performs useful tasks for humans or equipments excluding industrial automation application. Examples for personal use are domestic servant robot, automated wheelchair, personal mobility assist robot and pet exercising robot. Examples for public use are cleaning robot for public places, delivery robot in offices or hospitals, fire-fighting robot, rehabilitation robot and surgical robot in hospitals. In this context an operator is a person designated to start, monitor and stop the intended operation of a robot or a robot system.

7.1.2 机器人的定义

随着软件开发能力的发展，机器人的定义不断发生变化。以下定义适用于工业机器人。

- ISO：可自动被控制、可重新编程的多用途机械手，可编程控制三个轴或更多轴，可固定或移动，用于工业自动化应用。

- 美国机器人研究所（RIA）：可编程的多功能机械手，通过可编程的运动来移动材料、零件、工具或专用设备以执行各种任务。

- 英国机器人协会（BRA）：一种可重新编程的设备，通过可编程的运动来操纵和运输零件、工具或专用制造工具以执行特定的制造任务。

国际机器人联合会（IFR）对服务机器人进行了定义：

服务机器人是为人类或设备执行有用任务的机器人，不包括工业自动化应用。个人使用服务机器人包括家用仆人机器人、自动轮椅、个人移动辅助机器人和训练宠物机器人。公用服务机器人包括公共场所的清洁机器人、办公室或医院的送货机器人、消防机器人、康复机器人和医院里的手术机器人。这类机器人的操作员是指定启动、监视和停止机器人或机器人系统的指定人员。

7.2 Key Components of Industrial Robot

7.2 工业机器人的关键部件

Most industrial robots will comprise a robot arm, a controller, a monitor/terminal and a teach pendant. In newer robots the teach pendant and monitor/terminal are often combined.

大多数工业机器人由机械臂、控制器、监视器/终端和示教器组成。在新开发的机器人中，示教器和监视器/终端通常是组合在一起的。

7.2.1 Robot Arm

7.2.1 机械臂

Robot arms come in many configurations that are usually defined by the tasks to be carried out. Probably the most common configurations are the six-axis robot and the SCARA (selective compliance articulated robot arm). The main components of the arm are the drive motors and gear trains (or belts) that move the individual joints, and encoders that read the position of individual joints. Other components found on the arm are limit switches for joint travel limits, hard stops for the bigger joints, and power/data cable and control valves for end effectors. Less common forms of motive power include compressed air and hydraulics.

机械臂有多种结构，其通常由要执行的任务来确定。最常见的结构是六轴机器人和 SCARA（单柔性关节机械臂）。机械臂的主要部件是驱动电机和移动各个关节的齿轮系（或皮带）以及读取各个关节位置的编码器。臂上的其他组件包括用于限制关节行程的限位开关，用于较大关节的硬限位器以及用于末端执行器的电源/数据电缆和控制阀。不太常见的动力形式包括压缩空气和液压驱动。

The RX90L is a six-axis robot driven by electric motors (Fig. 7-1). It was developed for higher speeds and accuracy but with smaller payloads. The L in RX90L indicates that the robot arm is longer than the standard RX90. An end effector (tooling) such as a gripper, welding torch or spray gun is fitted to the end of the robot arm.

RX90L 是由电动机驱动的六轴机器人（图 7-1），它具有更高的速度和精度，但是其有效载荷较小。RX90L 中的 L 表示机械臂比标准的 RX90 长。在机械臂的末端安装有末端执行器（工具），如夹具、焊炬或喷枪。

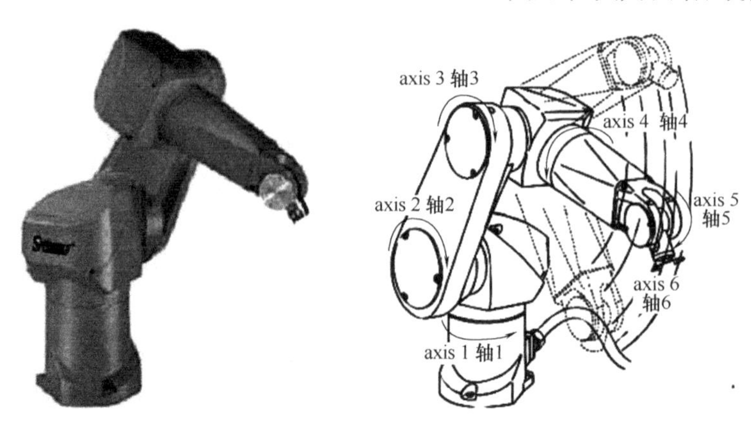

Fig. 7-1 RX90L six-axis robot

图 7-1 RX90L 六轴机器人

The Adept Cobra s600 robot is a four-axis SCARA robot (Fig. 7-2). There are three rotational joints and one translational joint. The rotational joints are one, two, and four. The translational joint is joint three.

Adept Cobra s600 机器人是一款四轴 SCARA 机器人（图 7-2），它有三个旋转关节和一个平移关节。关节 1、2、4 是旋转关节，关节 3 是平移关节。

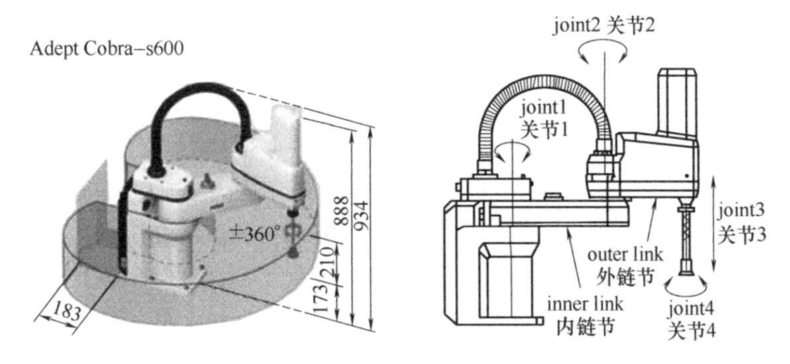

Fig. 7-2 Adept Cobra s600 Robot

图 7-2 Adept Cobra s600 机器人

7.2.2 Controller

The robot controller contains all the power and data elements necessary for the robot to carry out its function. A controller can usually be divided into three sections: power supply, power board and control/data board.

The power supply is usually a multi-tap transformer that converts three-phase or single-phase AC power into the required levels of DC or AC power. Motor drives will usually require 60 to 100 Volts while control and software boards will require 5 to 24 Volts. The controller will have many connectors, mostly at the back, the main ones being power for motor drives and data elements (encoders). All controllers will have some level of I/O and data ports (RS232/RS485).

The RX90L uses a CS7M controller which is shown in Fig. 7-3 (a). This controller is a modular controller and can be set up for different robots. There is a series of slot-in electronic boards at the front of the controller. These boards can be easily removed and changed with other boards for the purpose of upgrading or for a

7.2.2 控制器

机器人控制器包含机器人执行其功能所需的所有电源和数字元件。控制器通常可分为三个部分：电源、电源板和控制/数字板。

电源通过一个将三相或单相交流电转换为所需的直流或交流电的多级变压器供应。电动机通常需要 60～100V 的电压，而控制和软件板需要 5～24V 的电压。控制器有许多连接器，大多数位于控制器的背面，主要为电动机和数字元件（编码器）供电。所有控制器都有不同的 I/O 和数据端口（RS232/RS485）。

RX90L 使用 CS7M 控制器，如图 7-3（a）所示，该控制器是模块化控制器，可以对其进行设置以用于不同的机器人。控制器正面有一系列插槽式电子板，在为了升级或用于不同类型的机器人时，这

different type of robot. The control software on-board the CS7M is V+. This is the language used for programming the robot and it is very similar to using a computer language like Basic.

些电子板很容易与其他板一起被移除和更换。CS7M上的控制软件是V+，这是用于编程机器人的语言，其使用方法与Basic等计算机语言非常相似。

The Adept Cobra s-series robots require an Adept smart controller. The smart controller is used to control and program the robots, running on the Adept Smart Servo distributed motion control platform. The Adept Smart controller Cx has an integrated vision option and a conveyor tracking option, in addition to motion control [Fig. 7-3 (b)].

Adept Cobra s系列机器人需要Adept智能控制器。智能控制器用于控制和编程机器人，这种控制器在Adept Smart Servo分布式运动控制平台上运行。Adept Smart controller Cx除了具有运动控制功能外，还具有集成的视觉选项和传送带跟踪选项功能，如图7-3（b）所示。

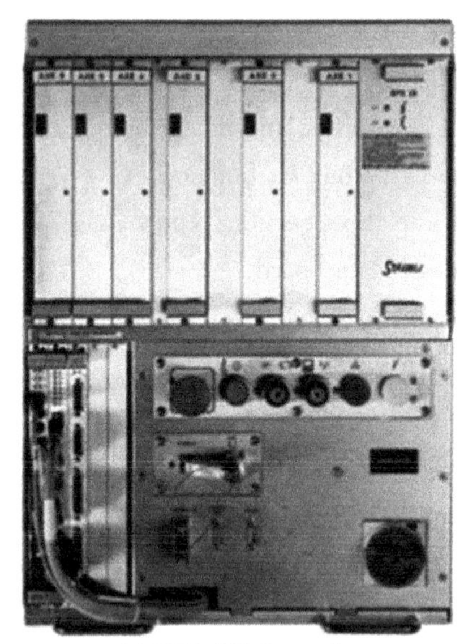

(a) RX90 CS7M controller
(a) RX90 CS7M控制器

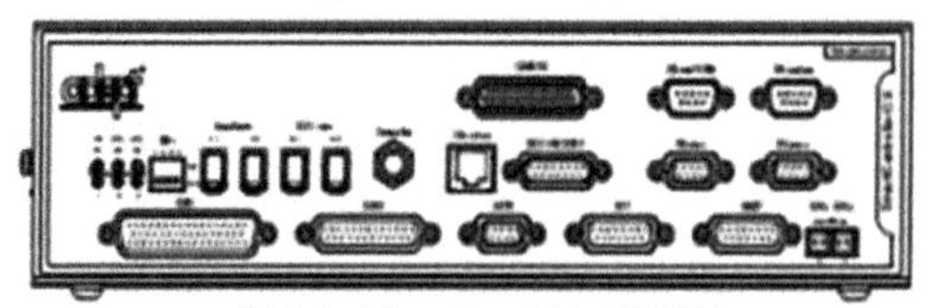

(b) Adept Smart contoller CX(10)
(b) Adept智能控制器CX(10)

Fig. 7-3 Controller
图 7-3 控制器

7. 2. 3 Teach Pendant

The function of a teach pendant, also called an MCP (manual control pendant), is to control the robot manually and to record the locations (points) for a particular program (Fig. 7-4). It can be used to move the robot and teach locations for storage by the controller and to select and step through programs. All teach pendants incorporate a dead man's mechanism so the pendant will not work without manual activation of a switch. The main buttons on a teach pendant are run/stop, joint selection, coordinate system section and speed level selection.

7. 2. 3 示教器

示教器，也称为 MCP（手动控制器），其功能是手动控制机器人并记录特定程序的位置（点）（图 7-4）。它可用于移动机器人，设定控制器的存储位置以及选择和逐步执行程序。所有的示教器都采用了自锁机制，如果没有手动启动开关，将无法工作。示教器上的主要按钮有运行/停止、关节选择、坐标系和速度级别选择。

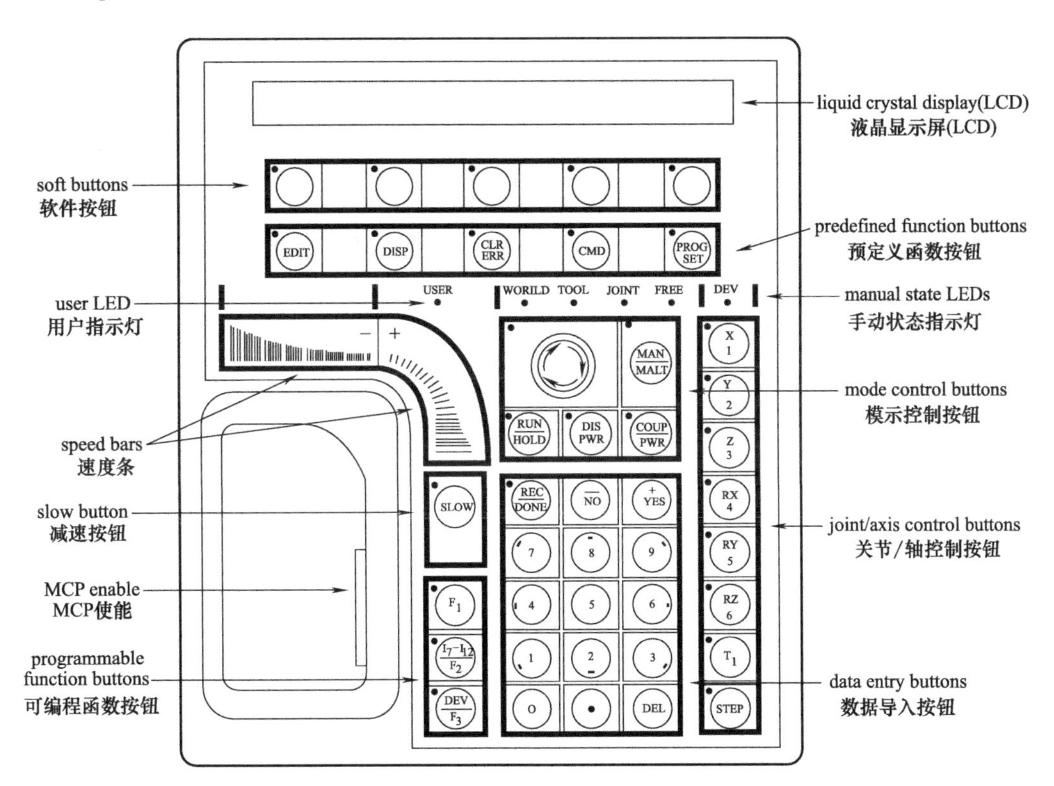

Fig. 7-4 Manual control pendant (MCP) layout

图 7-4 手动控制示教器（MCP）的布局

7. 2. 4 Monitor/Terminal

Programs are written and edited through some form of monitor/terminal. Newer robot systems will usually have

7. 2. 4 监视器/终端

通常，程序是通过监视器/终端以某种形式来编写和编辑的。新近推

a connector so that a PC, laptop or tablet can act as the terminal. The RX90L uses a WYSE terminal for monitoring, writing, and editing programs. The Adept Cobra s-series uses AIM software on a laptop.

出的机器人系统通常具有连接器，以便与PC、笔记本电脑或平板电脑等终端设备互连。RX90L机器人用WYSE终端来监视、编写和编辑程序；Adept Cobra s系列的机器人则是利用笔记本电脑上的AIM软件。

7.3 Robot Specification

7.3 机器人规格

The specifications given by a robot manufacturer allow the potential user to decide if the robot is suitable for the task in mind. Some of the common specs to be considered are the working envelope, payload, axis data, speed and repeatability, flexibility of installation, power consumption and ease of programming.

机器人制造商给出的规格能够使潜在用户确定机器人是否适合于某项任务。通常，需要考虑的一些常见的规格有工作范围、有效负载、坐标轴数据、速度和可重复性、安装灵活性、功耗和编程简易性。

The Staubli RX90L is a six-axis robot with work cell volume of approximately $9m^3$. It has a nominal payload of 3.5kg. The robot is driven by DC servo motors and has a repeatability of ± 0.025mm. The joint speed of the robot is approximately 12 m/s and it has a linear speed of 2.5m/s. The rated power is approximately 4kW. The robot can be floor, wall, or ceiling mounted.

Staubli RX90L是一款六轴机器人，其工作单元体积约为$9m^3$，标准有效载荷为3.5kg。这款机器人由直流伺服电机驱动，重复度为±0.025mm。此机器人的关节速度约为12m/s，线速度为2.5m/s，额定功率约为4kW。这种机器人可以安装在地板、墙壁或天花板上。

The Adept Cobra s600 has a reach of 600 mm and a payload of 2 kg (rated) and 5.5 kg (max).

- Joint range: joint 1 is ± 105°, joint 2 is ± 150°, joint 3 is 210 mm, joint 4 is ± 360°.
- Joint speeds: joint 1 is 386°/s, joint 2 is 720°/s, joint 3 is 1100 mm/s, joint 4 is 1200°/s.
- Repeatability: X axis and Y axis are ± 0.017 mm, Z axis is ± 0.003mm, θ is ± 0.019°.
- Digital I/O channels: 12 inputs, 8 outputs (included) and 4 solenoid outputs (end of arm).

Adept Cobra s600机器人的工作范围为600mm，有效载荷为2kg（额定）及5.5kg（最大）。

- 关节运动范围：关节1为±105°，关节2为±150°，关节3为210mm，关节4为±360°。
- 关节速度：关节1为386°/s，关节2为720°/s，关节3为1100mm/s，关节4为1200°/s。
- 可重复度：X、Y轴为±0.017mm，Z轴为±0.003mm，θ为±0.019°
- 数字I/O通道：12个输入、8个输出（包括）和4个螺线管输出（机械臂末端）。

7.4 Programming in V+

Both Staubli and Adept robots are programmed by V+. This is a computer control system and programming language designed for use by industrial robots. It consists of monitor commands and program instructions. Monitor commands are used to prepare the system to run a program written by the user. Program instructions allow the creation of programmers which can control the robot' s actions.

(1) Robot Pose

These are used to specify the destinations of robot motions. A distinction must be made between a "point" or "position" and a "pose". The first two terms are used to refer to a point in space defined by distinct x, y and z coordinates. A pose value refers to a point in the robot' s workspace and an orientation at that point and will have six values. There are two types of pose values:

- A precision point where the robot pose is represented by the exact position of the individual robot joints. It is a record of the degree of rotation of each of the six joints at the moment of recording and is signified by #. Precision point format: jt1, jt2, jt3, jt4, jt5, jt6 (degrees).

- A transformation is a robot independent representation of the position and orientation of the robot tool. Robot independence is achieved by defining pose in terms of a cartesian reference frame (x-y-z) fixed to the base of the robot (world frame). The position of the tool tip is defined by (x, y, z) coordinates and the tool orientation is defined by three angles [yaw, pitch and roll, (y, p, r)].

Transformation format: X Y Z (mm) y p r (degrees)

7.4 V+ 编程

Staubli 和 Adept 机器人都采用 V+进行编程。这是一种为工业机器人设计的计算机控制系统和编程语言。它由监视器指令和程序指令组成，其中监视器指令用于运行用户编写的程序，程序指令可以用于创建能够控制机器人动作的程序。

(1) 机器人位姿

机器人位姿用于明确机器人运动的终点。必须学会区分"点"或"位置"和"位姿"。前两个术语用于表示由不同的 x、y 和 z 坐标定义的空间中的点。位姿值表示的是机器人工作空间中的一个点以及该点所处的方位，并且其具有六个值。通常，有两种类型的位姿值：

- 机器人的位姿由机器人各个关节的精确位置来表示。它用于记录六个关节中每个关节在运动时的旋转角，并通过 # 来表示。精确的点的格式为：jt1、jt2、jt3、jt4、jt5、jt6 (度)。

- 转换是表示机器人器具位置和方向的特有方法。通过将笛卡尔坐标系 (x-y-z) 固定到机器人的基座 (世界坐标系) 上来定义位姿，从而实现机器人的独立性。器具尖端的位置由 (x,y,z) 来定义，其方位由三个角度即偏航角、俯仰角和滚动角 (y、p、r) 来定义。

转换格式：X Y Z (mm) y p r (度)

Joint configuration is not recorded and the arm will take the shortest route to the defined position. Transformations may allow more freedom of movement than precision points, however, they may include some rounding errors in motion calculations. The robot program can change a transformation by adding to or subtracting form the X, Y or Z value.

在不记录关节形状的情况下，机械臂将以最短的路径到达定义的位置。转换比精确的点有更多的运动自由度，然而，它们可能包括运动计算中的一些舍入误差。机器人程序可以通过加上或减去 X、Y 或 Z 值来改变一个转换。

When using the RX90L robot a precision pose is used at the start of the program in order to define the configuration of the robot arm, either "elbow up" or "elbow down" as shown in Fig. 7-5. A different robot pose can reach the same physical position. This not important for the SCARA type robot as the configuration of the robot arm is decided by the structure of the robot.

当使用 RX90L 机器人时，在程序开始时利用精确的位姿来定义机械臂的形状，"肘部向上"或"肘部向下"，如图 7-5 所示。不同机器人可以达到相同的身体姿势。由于机械臂的形状由机器人的结构来决定，因此这对于 SCARA 型的机器人来说并不重要。

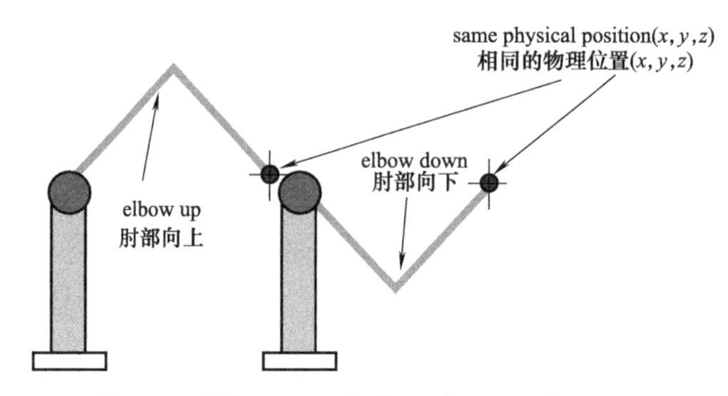

Fig. 7-5 Elbow up and elbow down configuration

图 7-5 肘部向上和肘部向下的形状

(2) Variable Name

(2) 变量名称

Variable names are assigned by the programmer. Examples of variables are poses, and counters. A variable name can have any number of characters. Each name must start with a letter and can only contain letters, numbers and full stops. Examples are PICK, PLACE, GRASP, Counter. The following are invalid: 3P, part-x.

变量名称由程序员分配，如令位姿和计数器为变量。变量名称可以是任意多个字符。每个变量名称必须以字母开头，并且只能包含字母、数字和句点。例如 PICK、PLACE、GRASP、Counter 为有效的变量名称，而 3P、part-x 则是无效的。

(3) Methods of Moving

(3) 移动的方法

• Joint interpolated moves: the robot moves only using the necessary joints to achieve the correct position. It's

• 关节插值运动：机器人仅使用必要的关节移动以获得正确的位

worth noting that all joint interpolated moves are curves.

置。值得注意的是，所有关节插值运动都为曲线。

• Straight Line moves：V+ initiates motion in a straight line which is the shortest distance between two points, normally all joints move. It's worth noting that some points cannot be programmed in straight line motion, i. e. , points passing through the robot body. Some common V+ codes are describle in Table 7-1.

• 直线运动：V+以直线的方式来运动，直线是两点之间的最短距离，通常所有关节都运动。值得注意的是，某些通过机器人本体的点不能以直线运动的方式来编程。一些常见的 V+代码见表 7-1。

Table 7-1　V+ codes

表 7-1　V+代码

V+指令	示例	说明
Move "location name"	Move PICK	移动抓手到 PICK
Moves "location name"	Moves PICK	以直线运动的方式移动抓手到 PICK
Depart "distance"	Depart 50	沿刀具 Z 轴从当前位置移动到 50mm 以外的点
Appro "location name", distance	Appro PLACE，75	沿刀具 Z 轴接近 PLACE，但停止在离它 75mm 远的地方
Appros "location name", distance	Appros PLACE，75	以直线运动的方式沿刀具 Z 轴接近 PLACE，但停止在离它 75mm 远的地方
Speed "%"	Speed 20	以正常速度的百分之二十来运动
Delay "seconds"	Delay 1	延时 1 秒
Type "…"		在屏幕上输入信息
Prompt "…", variable	Prompt "Input no of boxes", box	在屏幕上显示一条消息，要求操作员输入信息并将输入存储在变量中
Drive 'joint number', 'angle', 'speed'	Drive 6，35，10	以正常速度的百分之十来使关节 6 旋转 35°
Align		将刀具 Z 轴与最近的世界轴对齐
If 'signal'	If Sig（1029）	检查信号 1029 是否为 ON
Wait 'signal'	Wait Sig（1029）	暂停程序，直到收到特定信号为止
Set 'variable' = 'location name'	Set pick=pallet	为变量设置与位置托盘相同的坐标系
Shift (location name by x, y, z)	Shift (pick by 100，0，0)	将 PICK 的坐标改为（x 轴 100mm，y 轴 0，z 轴 0）
OPEN		下一次操作时打开抓手
CLOSE		下一次操作时关闭抓手
OPENI		打开抓手
CLOSEI		关闭抓手

7.5 Flow Diagram

7.5 流程图

The first step in creating a good program is to make an accurate flow diagram. This will show in block form how the program is to progress from start to end. Each step in the process is represented by a different symbol and contains a short description of the process step. The flow diagram symbols are linked together with arrows showing the process flow direction.

创建一个好程序的第一步是制作一个精确的流程图。流程图以块的形式显示程序从开始到结束的进度。该过程中的每个步骤都由不同的符号表示，并包含过程步骤的简短描述。流程图的符号通过箭头连接在一起，并显示出进程的方向。

A good program contains all the following:

- A suitable remark statement explaining the object of the program, the program name, date, revision number, etc.
- A precision point move to get correct arm orientation.
- An area near the start of the program where all variables are set to an initial value.
- Sensible names for all user variables including program name.

一个好的程序包含以下内容：

- 恰当的备注声明，解释程序的目的、程序名称、日期、修订号等。
- 一个精确点移动，能够得到正确的手臂方位。
- 在程序开头附近的区域为所有变量都设置了初始值。
- 为所有用户变量包括程序名称都定义了合理的名称。

The flow diagram symbols are proposed as Fig. 7-6.

流程图的符号见图 7-6。

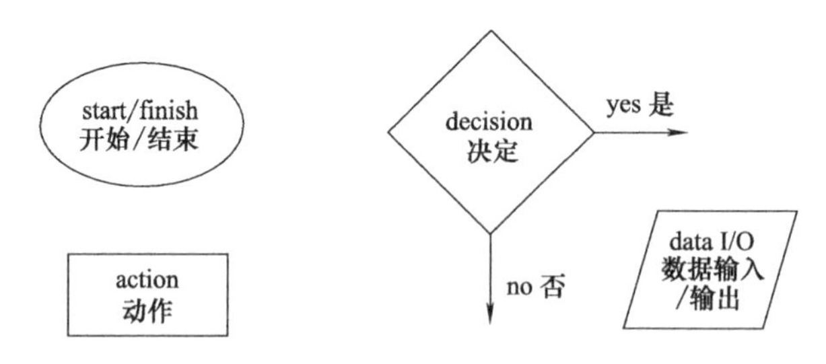

Fig. 7-6 Flow diagram symbols

图 7-6 流程图符号

7.6 Robot Program

7.6 机器人程序

The three examples below demonstrate the control structures used in algorithms: sequential operation, conditional operation and iterative loop.

下面的三个示例说明了算法中使用的控制结构：顺序操作、条件操作和迭代循环。

【Example 7-1】 A robot is used to draw a straight line.

【例 7-1】 用机器人来绘制直线。

The robot moves into position, moves along in a straight line and then returns to its home position. The flow diagram is shown in Fig. 7-7.

机器人移动到位，沿直线移动然后返回其原始位置。流程图如图7-7所示。

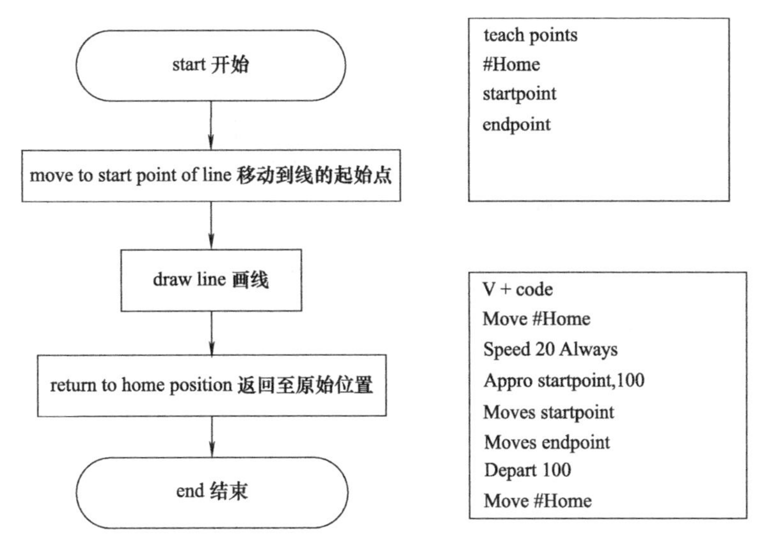

Fig. 7-7 Figure for example 7-1

图 7-7 例 7-1 附图

The robot does each movement in sequence and returns to its home position when it is finished.

机器人按顺序完成每个动作，并在完成后返回其原始位置。

【Example 7-2】 If a part is in place a sensor will be ON. If the sensor is ON the robot will move and pick up the part. If the sensor is not ON the robot will move to the home position. The flow diagram is shown in Fig. 7-8.

【例 7-2】 如果零件在正确的位置，则传感器就输出 ON。如果传感器输出为 ON，机器人将移动并拾取零件。如果传感器输出不为 ON，机器人将移动到原始位置。流程图如图 7-8 所示。

This example introduces the conditional control structure. The diamond shape in the flow diagram indicates that a question is being asked. In this case the question is whether the sensor is ON. If the answer is yes then the robot will move to pick up and place the part. The robot then moves to its home position and the program ends. If the answer is no then the robot moves directly to the home position without picking up the part and the program ends. When drawing the diamond symbol in a flow diagram it is important to show both the yes and no options.

此示例介绍的是条件控制结构。流程图中的菱形表示判断条件。在这种情况下，判断的条件是传感器是否为 ON。如果是，则机器人将移动进而拾取并放置零件，然后机器人移动到其原始位置，程序结束。如果不是，则机器人直接移动到原始位置而不拾取零件，程序结束。在流程图中绘制菱形符号时，同时写出是和否选项是很重要的。

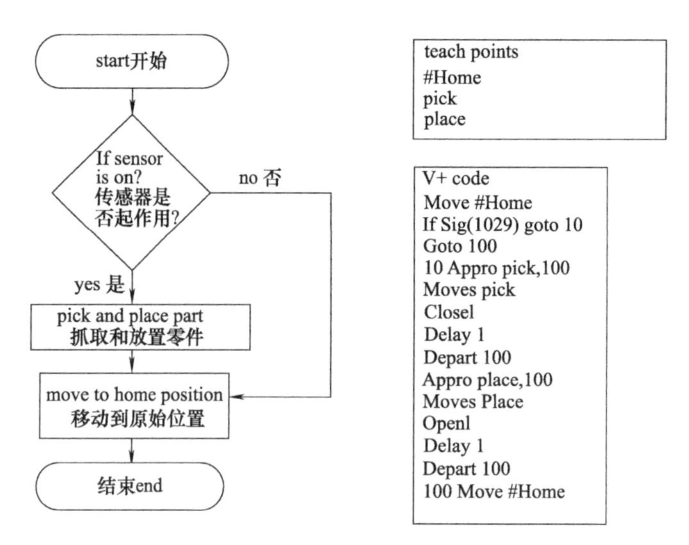

Fig. 7-8 Figure for example 7-2

图 7-8 例 7-2 附图

【Example 7-3】 In iterative operation a robot moves 100 parts from one conveyor to another. When the task is complete the robot moves to home position. The flow chart can be seen in Fig. 7-9.

【例 7-3】 在迭代操作中，机器人从一个传送机移动 100 个零件到另一个传送机，任务完成后，机器人移动到其原始位置。流程图如图 7-9。

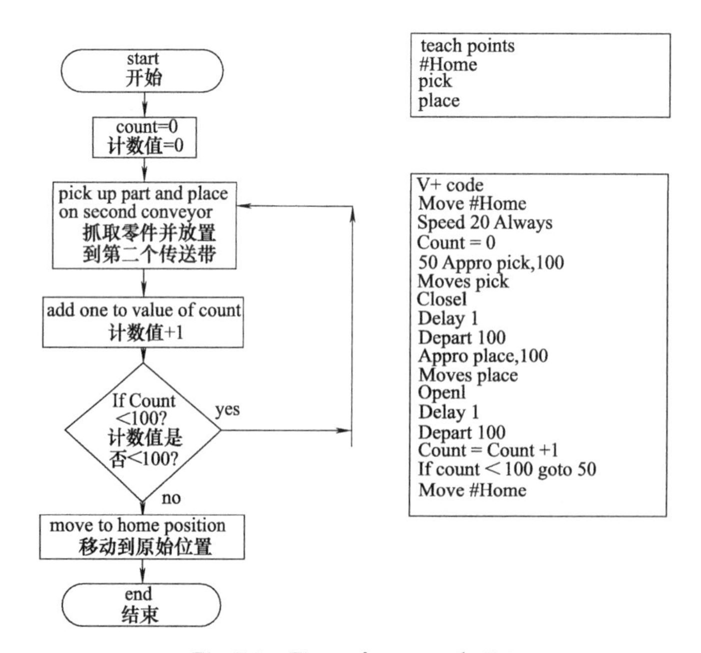

Fig. 7-9 Figure for example 7-3

图 7-9 例 7-3 附图

The value of the variable "count" is put equal to 0. The robot then picks a part up from the first conveyor and places it on the second conveyor. 1 is then added to the value of the variable "count". The robot then checks if the value of "count" is less than 100. If the answer is yes the robot goes back and picks and places another part. The value of "count" is increased by 1 each time a part is picked and placed. When 100 parts have been picked and placed the value of "count" will be 100 so the answer to the question as to whether the count is less than 100 will be no. The robot will move to the home position and the program will end.

首先，将变量“count”的值置为0。然后，机器人从第一个传送机上拾取一个零件并将其放在第二个输送机上，随后将1赋值给变量“count”。机器人判断“count”的值是否小于100。如果是，则机器人返回去拾取并放置另一个零件。每次拾取和放置一个零件，“count”的值就增加1。当拾取并放置100个零件时，“count”的值将为100，此时，条件“count是否小于100”的判断为否，因此机器人将移动到原始位置，程序结束。

Fig. 7-10 shows how a number of loop structures can be used in V+program code.

图7-10给出了如何在V+程序代码中使用多个循环结构。

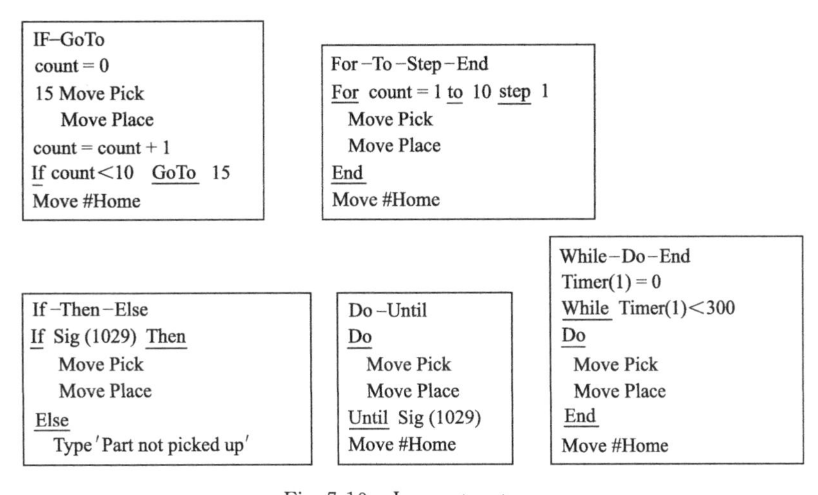

Fig. 7-10 Loop structures

图7-10 循环结构

• If _ GoTo is the loop structure used in example 7-3 above.

• If _ GoTo是例7-3中使用的循环结构。

• For _ To _ Step _ End allows us to set up a loop where we give the range for the variable to count through. Step shows us how to count. If step is 1 we count every time, if step is 2 we count every second time. End indicates where the loop stops.

• For _ To _ Step _ End使我们能够设置一个循环，并给出变量的计数范围。step使我们能够看出如何计数。如果step为1则每次都计数，如果step为2则每隔一次计数。End表示循环停止的位置。

• If _ Then _ Else is a conditional statement with two options.

• If _ Then _ Else 表示带有两个选项的条件语句。

• Do _ Until carries out the command sequence until something happens to stop it, in this case when Sig (1029) comes on.

• Do _ Until 表示执行命令序列，直到某件事情发生则停止，对于图示这种情况来说，是当 Sig (1029) 启动时。

• While _ Do _ End sets up a loop that should continue while something is happening, in this case while the current value of the timer is less than 3 seconds.

• While _ Do _ End 表示设置一个循环，当某件事发生时应该继续，对于图示这种情况来说，是当计时器的当前值小于 3 秒时。

Chapter 8　Vision System
第8章　视觉系统

Key words 重点词汇

image acquisition	图像采集	image enhancement	图像增强
lighting	照明	image analysis	图像分析
pixel	像素	high pass filter	高通滤波器
resolution	分辨率	low pass（mean）filter	低通（均值）滤波器
image digitization	图像数字化	median filter	中值滤波器
gray scale image	灰度图像	binary image	二值图像
image processing	图像处理	thresholding	阈值化
image correction	图像修正	calibration	校准，标定

8.1　Introduction of Vision System

8.1　视觉系统简介

A vision system involves the acquisition of image data, followed by the processing and interpretation of this data by computer for some useful application.

视觉系统涉及图像数据的采集，通过计算机对这些数据进行处理并进行解释说明，进而将其运用到实际中。

Fig. 8-1 shows a brake disc from a car and an image from the vision system used to inspect the part. The image indicates that the part is ok and gives values for the area and the centre of gravity.

图 8-1 为汽车的制动盘和用于检查零件的视觉系统中的图像。图像显示零件没问题，并给出了区域和重心的一些值。

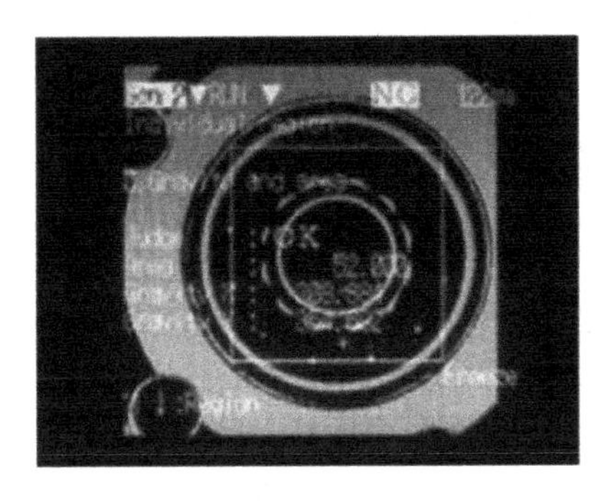

Fig. 8-1　Actual brake disc and screen shot from vision system

图 8-1　实际制动盘和视觉系统的屏幕截图

8.1.1 Uses of Vision Systems

- Verification of presence of a component.
- Part inspection.
- Dimensional measurement/gauging.
- Process adjustment/control.
- Motion control.
- Quality control.
- Object recognition.

8.1.1 视觉系统的用途

- 检查零部件是否存在。
- 零件检查。
- 尺寸测量/估计。
- 过程调整/控制。
- 运动控制。
- 质量控制。
- 物体识别。

8.1.2 Outputs of Vision Systems

The output from a vision system can be in the form of a classification or a measurement. Classification indicates if the part is present, if the part is the correct shape, and if it has the correct features, while measurement outputs the properties of the part such as diameter and centre of gravity. The application will dictate the outputs required, the image processing techniques and hardware to be used.

8.1.2 视觉系统的输出

视觉系统可以用分类或测量的形式进行输出。分类用以表明零件是否存在，零件形状是否正确或零件是否具有正确的特征。测量输出零件的性质，如直径、重心。实际应用决定所需要的输出、图像处理技术和硬件。

8.1.3 Components of Vision System

The components of a vision system, as shown in Fig. 8-2, are lighting, camera, digitized hardware and software for image processing.

8.1.3 视觉系统的组成

视觉系统的组成如图 8-2 所示，包括用于图像处理的照明、相机、数字化硬件和软件。

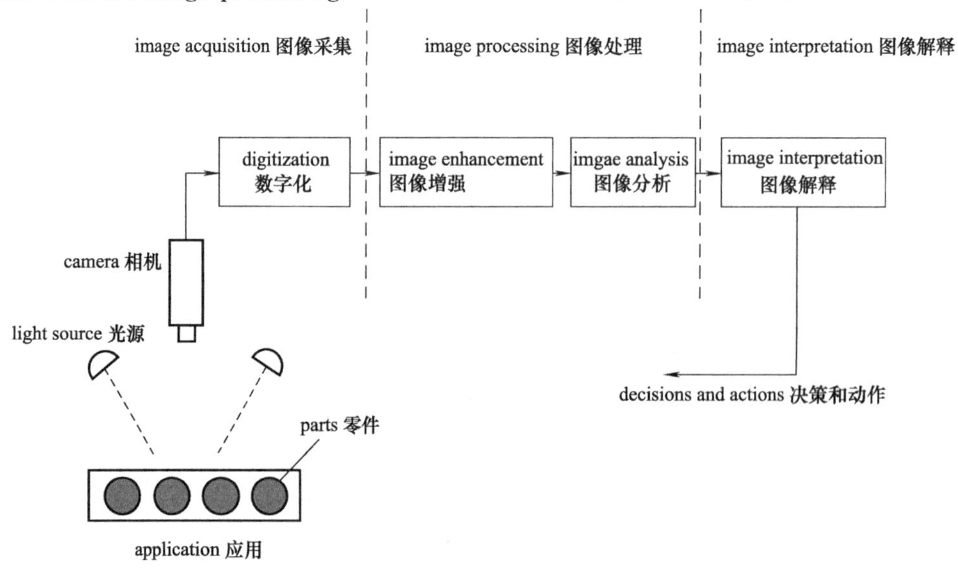

Fig. 8-2 Vision system components

图 8-2 视觉系统的组成

8.2 Image Acquisition

Acquisition of a clear image relies on good lighting and positioning of the component. It is essential to use lighting so that ambient lighting conditions do not cast a shadow or glare on the part to be inspected. Consistent lighting makes image processing easier and less time consuming. In addition, the image processing can be clearer by looking for the appropriate positioning when positioning the components.

8.2 图像采集

清晰图像的采集取决于良好的采光和部件的定位。因此，必须有效采光以便周围的照明条件不会对要检查的部件产生阴影或眩光等。一致的采光使图像处理更容易、耗时更少。此外，在定位部件时寻找恰当的定位，能够使图像处理更清晰。

8.2.1 Light Source

Typical light sources are florescent lamp, halogen lamp and LED. They come in different shapes and colours and are chosen based on the object position and colour (Fig. 8-3).

8.2.1 光源

典型的光源有荧光灯、卤素灯和LED。它们具有不同的形状和颜色，通常需要根据物体位置和颜色来选择光源（图 8-3）。

Fig. 8-3 Light sources

图 8-3 光源

Ring lights are used to provide diffused illumination over a small area directly in front of the camera. Area lights give concentrated illumination that allows a camera to detect the presence or absence of a feature. Back lights and high-power back lights give an even, low intensity illumination that creates a silhouette ideal for inspecting correct size and shape. Liner array lights are used for high-intensity illumination of larger areas at long distances. On-axis lights provide even, diffused illumination of flat, reflective surfaces while spot

环形光源用于在相机正前方的小区域直接进行漫射照明。区域灯用于集中照明，使相机可以检测某个特征是否存在。背光灯和大功率背光灯提供一个均匀、低强度的照明，产生适合尺寸和形状检查的轮廓。线性阵列灯用于远距离大面积的高强度照明。同轴灯用于平坦、反光表面的均匀漫射照明。聚光灯用于在视觉传感

lights offer very bright illumination in vision sensor applications. The laser structure lamps are used to detect small objects in precise positioning control.

器场合中非常明亮的照明。激光结构灯用于检测精确定位控制中的小物体。

Table 8-1 gives the benefits and drawbacks of different coloured lights and suggests some typical applications.

表 8-1 给出了不同颜色的灯的优点和缺点，并对不同颜色的灯给出了建议的应用场合。

Table 8-1 Colours used in lighting for vision

表 8-1 用于视觉系统灯光的颜色

colour 颜色	white 白色	infrared 红外线(IR)	red 红色	green 绿色	blue 蓝色	ultraviolet 紫外线(UV)
wavelength 波长	390～780nm WHI	850～1050nm	625～660nm	530nm	470nm	365～395nm
benefits 优点	every wavelength of spectrum 具有频谱的每个波长	invisible 不可见	least expensive 价格低		high scattering rate 散射率高	365nm light can cause materials to flouresce 365nm 的光可使材料产生荧光
drawbacks 缺点	difficult to avoid interference from ambient light 难以避免周围环境中光线的干扰	requires special polarization filters and IR lens 需要特殊的偏振滤镜和 IR 镜头			can create noise on an object with many defects 可对具有许多缺陷的对象产生噪声	safety precautions 需要做好安全防护措施
typical applications 典型应用	colour camera 彩色相机	where an object needs to be made see through 用于对象需要透视的场合	any 任何场合	for inspecting green coloured objects such as PCBs or wafers 用于检查 PCB 板或晶片等绿色物体	for applications requiring higher resolution of minor defects 适用于针对小缺陷需要较高分辨率的场合	for inspecting glue, adhesive paint, UV ink 用于检查胶水、黏合剂、油漆、UV 油墨

As shown in Fig. 8-4, various lighting arrangements are used to emphasise the features that are being inspected in a particular part. They can be broadly classified as front lighting and back lighting. Front lighting is used where it is important to see surface features while back lighting is used where the profile of the part and any holes in the part need to be inspected.

如图 8-4 所示，通过不同的灯光布置来突出某个零件需要检查的部分的特点。大致可分为正面照明和背面照明（逆光照明），其中正面照明用于需要观察物体表面特征的场合；背面照明用于需要检查零件轮廓和零件中任意孔的位置的场合。

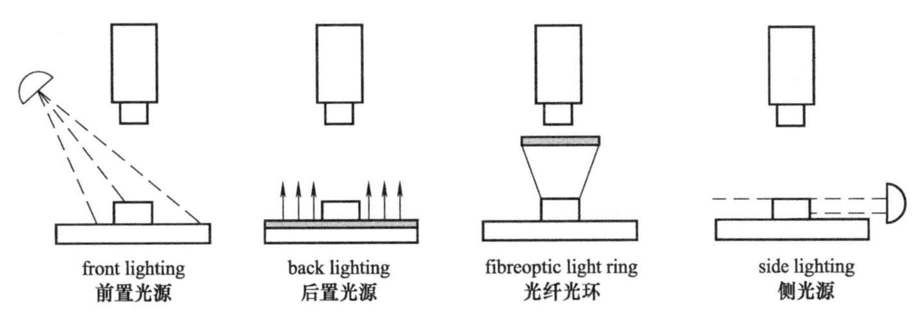

Fig. 8-4 Lighting arrangements

图 8-4 灯光的布置

8.2.2 Camera

8.2.2 相机

Two camera types are used in machine vision, CCD and CMOS. A CCD camera has a charge coupled device—photon detector. It consists of a thin silicon wafer divided into a regular geometric array of thousands or millions of light-sensitive regions. This capture and stores image information in the form of localized electrical charge that varies with incident light intensity. The variable electronic signal associated with each picture element (pixel) of the detector is read out very rapidly and used as an intensity value for the corresponding image location.

在机器视觉中，通常使用两种类型的相机：CCD 和 CMOS. 。CCD 相机有一个电荷耦合元件——光子探测器。它由一个薄硅晶片组成，该晶片被分成数千或数百万个感光区，并形成几何阵列。它能够以局部电荷的形式捕获和存储图像信息，并且随着入射光的强度而变化。进而使与检测器的每个像素（像素）有关的可变电子信号被非常快速地读出来，并作为对应图像位置的强度值。

Following digitization of these values, the image can be reconstructed and displayed on a computer monitor virtually instantaneously.

在对这些值进行数字化之后，可以重建此图像并在计算机显示器上实时显示出来。

A CMOS (complementory metal-oxide semiconductor) sensor consists of an array of tiny cells which convert photon energy to electron. The cells are surrounded by transistors which amplify the charge which is then translated into a particular colour value.

CMOS（互补金属-氧化物半导体）传感器由一组微小的电池组成，这些电池将光子能量转化为电子。电池被晶体管包围，这些晶体管将电荷放大，然后将电荷转换成特定的颜色值。

A pixel (short for picture element) is the smallest unit of a digital image or graphic that can be displayed

像素（图像元素的简称）是数字图像或图形的最小单位：可以在数

and represented on a digital display device. Pixels can be combined to form a complete image, video, text or other visible object on a computer moniter.

字显示设备上显示和表示。将像素组合在一起可以在计算机显示器上形成完整的图像、视频、文本或其他可见的东西。

The following relevant contents should be noted:

- Resolution of system is dependent on the number of pixels.
- Cost of camera increases with the number of pixels.
- Processing time increases with the number of pixels as the processor must digitize image from each pixel.
- Each pixel on a CMOS sensor has several transistors located next to it.
- CCD sensors create high quality, low noise images. CMOS sensors are more susceptible to noise.
- CCD uses a process that consumes up to 100 times more power than a CMOS sensor which is low power.
- CMOS chips can be fabricated on a standard silicon production line, so they tend to be extremely inexpensive compared to CCD sensors.
- CCD sensors have been mass produced for a long period, so they tend to be very reliable.

需要注意以下相关内容：

- 系统分辨率取决于像素数量。
- 相机的成本随像素数量的增加而增加。
- 由于处理器必须对图片的每个像素进行处理以对其进行数字化，因此处理时间随着像素数量的增加而增加。
- CMOS 传感器上的每个像素旁边都有几个晶体管。
- CCD 传感器可以创建高质量、低噪声的图像。一般来说，CMOS 传感器更容易受到噪声的影响。
- CCD 使用的过程比低功耗的 CMOS 传感器消耗的能量多 100 倍。
- CMOS 芯片可以在任何标准硅生产线上制造，因此与 CCD 传感器相比，它们的价格非常便宜。
- CCD 传感器已经大量生产了很长一段时间，所以它们非常可靠。

8.3 Image Processing

8.3 图像处理

8.3.1 Image Digitization

8.3.1 图像数字化

The image obtained from the camera is in the form of an analogue video signal which is converted to a digital signal where brightness values are assigned to each pixel. This method of processeing images is called digitization.

通常从相机中获得的图像是模拟视频信号的形式，该信号被转换为数字信号，其中亮度值被分配给每个像素。这种处理图像的方式称为数字化。

Generally when the camera captures an image, it is a colour image. Each pixel is represented by three pieces of information, RGB, i.e., the amount of red, green

通常，相机捕获的图像为彩色图像。每个像素由三个信息 RGB 来表示，即图像中的红色、绿色和蓝

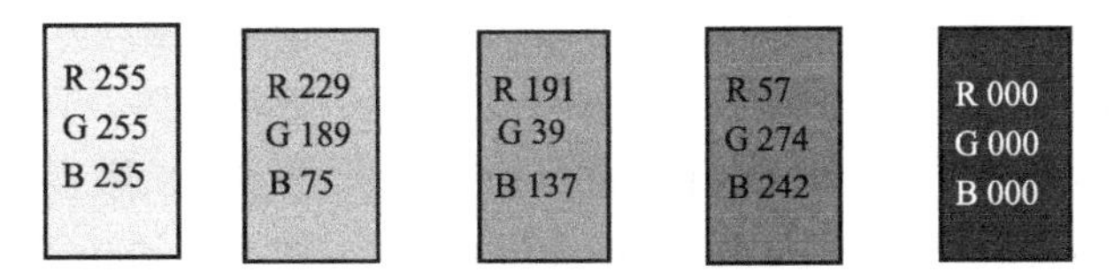

Fig. 8-5 RGB—numbers identifying Red, Green and Blue level 0 to 255
图 8-5 RGB——识别红色、绿色和蓝色的数字从 0 到 255

and blue in the image. Fig. 8-5 shows that each colour can be represented by RGB. The first is white which has the highest level of red, green and blue, 255. While black has the lowest level of red, green and blue, 0.

色的量。图 8-5 中每种颜色都可以用 RGB 来表示。其中白色具有最高数量的红色、绿色和蓝色，其值为 255。而黑色具有最低数量的红色、绿色和蓝色，其值为 0。

The values of red, green and blue are represented by 8 bits (1 byte) of information, so in order to represent one colour in a pixel requires 3 bytes. If using a 12millions pixel camera, it means that a single image will require 36 megabytes. The objective of the vision system is to extract meaningful data and in an industrial context, this can be done by using a much simpler form of the image which is considerably faster to process.

表示红色、绿色和蓝色的值各由 8 位（1 字节）信息来表示。因此，在像素中表示一种颜色需要 3 个字节。如果使用 1200 万像素的相机，这意味着单个图像的大小为 36 兆字节。视觉系统的目标是提取有意义的数据，由于处理简单的图像速度比较快，因此在工业环境中往往利用简单的图像。

The colour image is first converted to a grey scale image. This means that each pixel is represented by 8 bits of information, as shown in Fig. 8-6.

首先将彩色图像转换为灰度图像。其中每个像素由 8 位信息来表示，如图 8-6 所示。

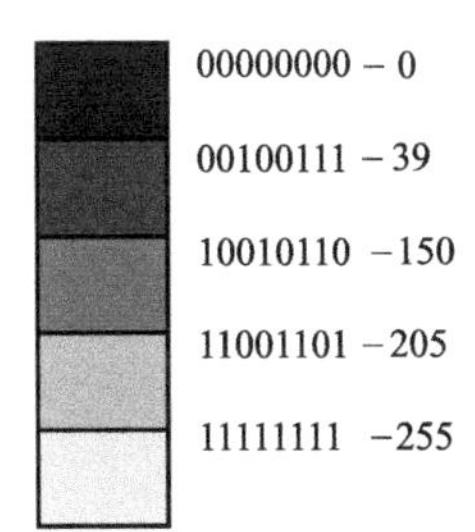

Fig. 8-6 Grey scale system, pixel brightness expressed by 256 levels
图 8-6 灰度系统，像素亮度分为 256 个等级

When an image is represented as a grey scale image, each pixel will have a value between 0 and 255 which represents a particular shade of grey. In Fig. 8-7 (a), four objects have been placed on a grid and an image taken.

当图像被表示为灰度图像时，每个像素将具有 0～255 之间的值，其表示特定的灰色阴影。在图 8-7(a) 中，将四个对象放置在网格上，并为它们拍摄图像。

Fig. 8-7(b) shows the digital matrix which represents this image in grey scale. The resolution in this case is 8×8 pixels. Each number represents the amount of light falling on the pixel so when the black square fully covers a pixel the grey scale value will be 0. If there is no object on a pixel it will have a value of 255 (white). The black square also partially covers 12 pixels. The corner values are 220 as less of the pixel is covered by the black object and 8 other edge pixels are 140.

图 8-7(b) 为以灰度表示该图像的数字矩阵。在这种情况下，分辨率为 8×8 像素。每个数字代表落在像素上的光子的量。因此，在黑方块完全覆盖像素时，其灰度值为 0。如果像素上没有对象，则其值为 255（白色）。同时，黑方块也不完全地覆盖了 12 个像素。由于在拐角的地方像素被黑色物体覆盖的面积较小，因此其值为 220，而其他八个边缘部分的像素为 140。

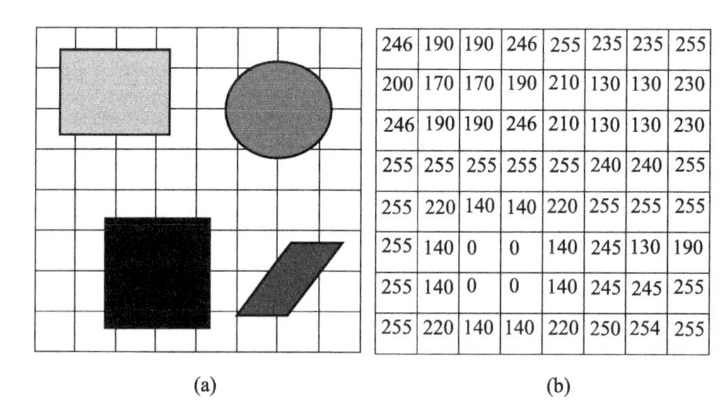

246	190	190	246	255	235	235	255
200	170	170	190	210	130	130	230
246	190	190	246	210	130	130	230
255	255	255	255	255	240	240	255
255	220	140	140	220	255	255	255
255	140	0	0	140	245	130	190
255	140	0	0	140	245	245	255
255	220	140	140	220	250	254	255

(a) (b)

Fig. 8-7 Grey scale image and digital matrix

图 8-7 灰度图像和数字矩阵

8. 3. 2 Stages of Image Processing

8. 3. 2 图像处理的步骤

Image processing is used to extract the required image features, e. g. , shape, orientation, dimension, from the raw image data. There are four stages as follows.

图像处理用于从原始图像数据中提取所需要的图像特征，例如形状、方向、尺寸等。图像处理分为以下四个阶段。

(1) Image Correction

(1) 图像校正

This is required in the pre-processing stage if image degradation occurs. There are two ways to deal with image degradation, i. e. , noise reduction and blur. The intensity value of each pixel is affected independently in the process of noise reduction. The grey level of individual pixel is changed due, for example, to electrical noise. Blur affects a number of pixels. Typical fac-

如果发生图像退化现象，则在预处理阶段需要进行图像校正。有两种方式来处理图像退化，即降噪和模糊。在降噪过程中每个像素的强度值独立地受影响。单个像素的灰度级会受到如电噪声的影响而改变。在模糊过程中，多个

tors such as poorly focused optics and object motion produce blurred image and geometric distortion, reasulting in a focused distorted imagine.

图像点的强度水平会同时受到影响。典型的因素，比如聚焦不良的光学和物体运动，都会产生模糊的图像和几何失真，从而产生聚焦失真的图像。

It is important to identify the cause of the degradation as it may be easier to correct imagine physically. For example, it's easier to correct by changing the alignment way of the light source and the camera than by expensive software and spending a lot of processing time.

由于通过改变物理条件更容易来校正图像，因此识别图像退化的原因是很重要的。例如，改变光源和相机的对准方式比使用昂贵的软件和花费大量处理时间更容易对其进行校正。

(2) Image Enhancement

(2) 图像增强

Image enhancemet can transform image to make it suitable for further processing. It enhaces visual appeal by improving image date, in particular in cases where a histogram of the pixel intensity values shows that the spread of grey levels used in the image is low. Each original grey level is assigned a new value when image date is modified. In general when all grey scale images show even and horizontal histogram distribution, the human eye can find the image elements are clearly distinguished.

图像增强可以改善图像，使其适合进一步处理。它通过改善图像的数据以增强视觉的吸引力，特别是在像素强度值的直方图显示图像的灰度分布较低的情况下。当图像数据被修改时，每个原始灰度级被分配一个新的值。一般情况下，当灰度图片中全部呈现出均匀且水平分布的直方图时，人眼可以发现图像元素清晰可辨。

(3) Image Transformation or Filtering

(3) 图像转换或滤波

Image transformation or filtering can extract useful information from an image. It helps to enhance the image digitally by removing unwanted elements due to noise or glare, or by emphasizing particular aspects of the image (such as an edge or contour). This process is called digital filtering and involves the application of a computational filter (or mask) to the numerical representation of the image. Filters can be used to smooth an image or to sharpen an image.

图像转换或滤波是为了从图像中提取有用信息，进而有助于通过去除噪声或眩光影响的元素或通过强调图像的特定方面（例如边缘或轮廓）来数字化地增强图像。该过程称为数字滤波，它涉及将计算滤波器（或掩膜）应用于图像的数字表示。滤波器用于平滑或锐化图像。

High pass filter Fig. 8-8 shows a typical high pass filter which is used to emphasize an edge in an image. The filter can also be called a kernel or a mask. The filter is applied

高通滤波器 图 8-8 为一个典型的高通滤波器，用于强调图像的边缘部分。这种滤波器也可以称为内核

to the image in Fig. 8-9 (a) by multiplying it across the im-age matrix from the top left. The result is placed in the centre pixel in the filtered image [Fig. 8-9 (b)].

或掩膜。高通滤波器应用于图 8-9(a) 中的图像，从图像矩阵的左上角开始依次乘以该滤波器，并将计算结果放于滤波后的图像的中心像素中［图 8-9(b)］。

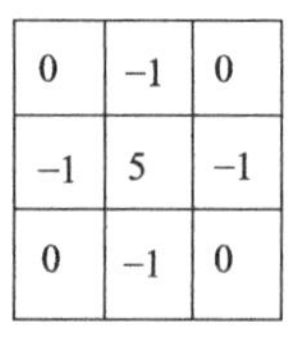

0	−1	0
−1	5	−1
0	−1	0

Fig. 8-8 High pass filter

图 8-8 高通滤波器

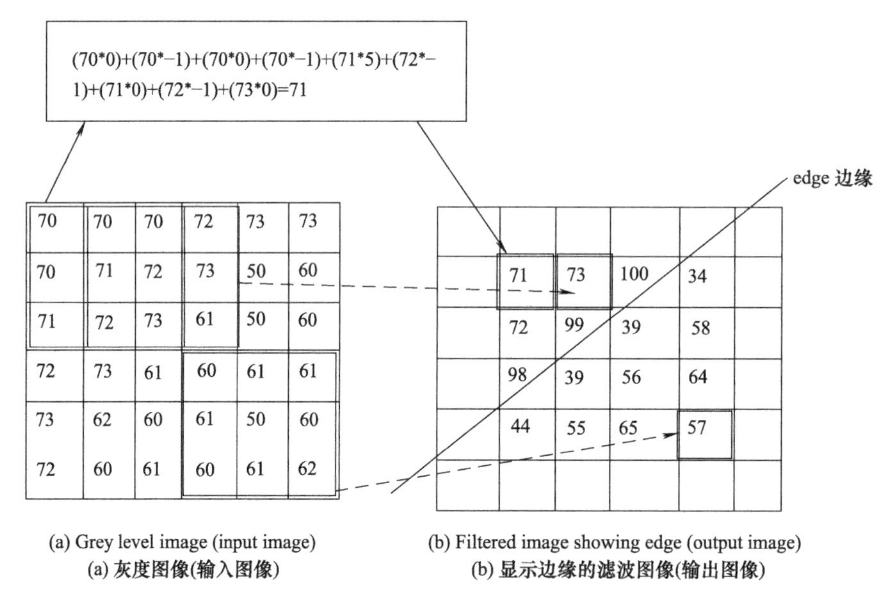

(a) Grey level image (input image)
(a) 灰度图像(输入图像)

(b) Filtered image showing edge (output image)
(b) 显示边缘的滤波图像(输出图像)

Fig. 8-9 Figure of high pass filter

图 8-9 高通滤波器图像

The results show that there is a larger difference be-tween the pixel values along the edge of the part.

结果表明，沿着零件边缘的像素值发生了大的变化。

Low pass (mean) Filter A low pass filter is used to find the mean value of a group of neighbours and replace the cen-tre value of this group with the mean value. It eliminates the pixel values that can not represent their surroundings Fig. 8-10 shows a low pass filter. This has the effect of removing point areas of light or dark. This is applied in the same way as the high pass filter and results in a smoothed (blurred) image, as shown in Fig. 8-11.

低通（均值）滤波器 低通滤波用它邻近的以及它自己的平均值来替换图像中心位置的像素值。它具有消除不能代表其周围环境的像素值的效果。图 8-10 为一个低通滤波器。它具有去除亮或暗的点区域的效果。其与高通滤波器的应用方式相同，并产生平滑（模糊）的图像，如图 8-11 所示。

1	1	1
1	1	1
1	1	1

Fig. 8-10 Low pass filter

图 8-10 低通滤波器

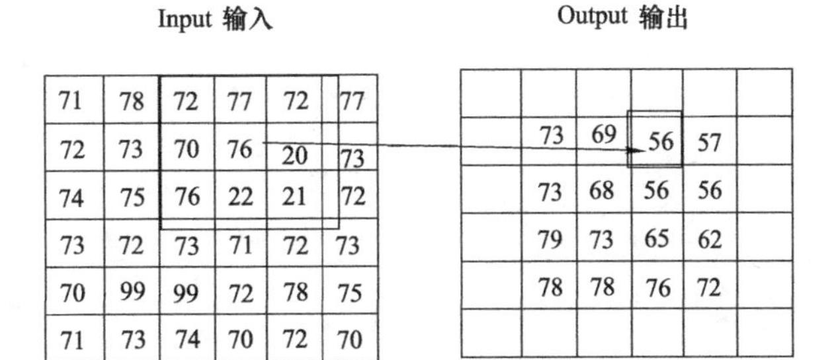

Fig. 8-11 Figure of low pass filter

图 8-11 低通滤波器图像

Median filter The median filter considers each pixel in the image in turn and looks at the pixel nearby to decide whether it represents its surroundings. In this case the pixel value is replaced by the median value rather than the mean, as shown in Fig. 8-12.

中值滤波器 与低通滤波器一样，中值滤波器将依次考虑图像中的每个像素并查看其附近的像素以确定它是否代表其周围环境。它不是简单地用相邻像素值的平均值替换该像素值，而是用这些值的中值来替换它，如图 8-12 所示。

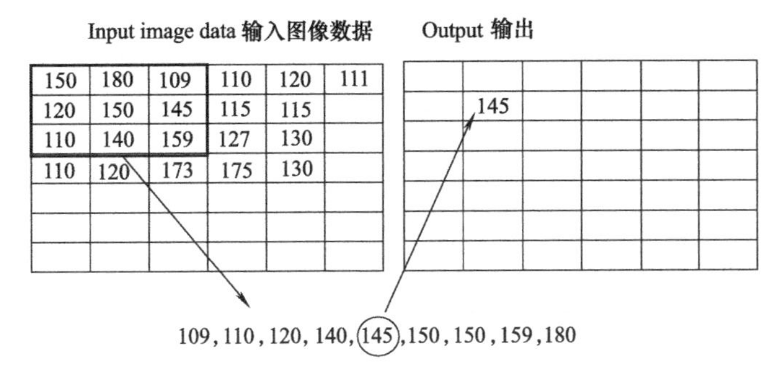

Fig. 8-12 Median filtering

图 8-12 中值滤波器

Using the median filter means that a very high or low value will not have a significant effect on the result, and as the median value is one value nearby, a new unrepresentative value resulting from calculations across an edge is not created. This means that the median filter outputs a sharper edge than the low pass filter, as shown in Fig. 8-13.

使用中值滤波器意味着一个非常高或很低的值不会对结果产生显著的影响。由于中值实际上是相邻的某个像素的值，当滤波器经过边缘时，不会产生新的不具代表性的像素值。因此中值滤波器在保留锐边方面要比低通滤波器好得多，如图 8-13 所示。

(4) Image Analysis

(4) 图像分析

It is used to produce the object characteristics and

图像分析用来产生对象的特征和测

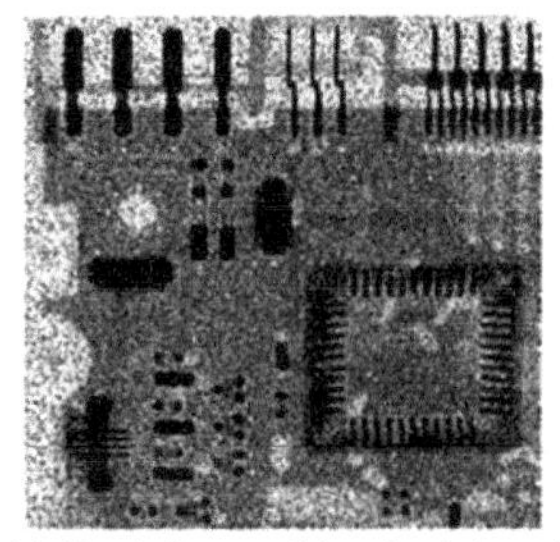

(a) *X*-axis image of the circuit board

(a) 电路板的*X*轴图像

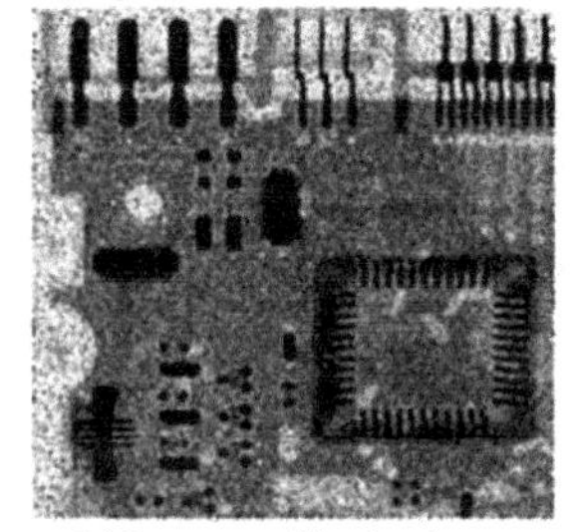

(b) Low-pass filtering image

(b) 低通滤波图像

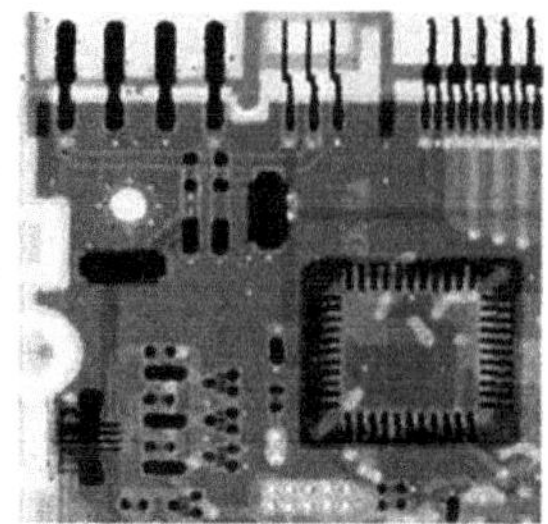

(c) Median filtering image

(c) 中值滤波图像

Fig. 8-13 Median filter output instance

图 8-13 中值滤波器输出实例

measurements. Steps are as follows.

量值，步骤如下。

Windowing This is a means of concentrating analysis on a small field of view, thereby conserving computer resources of run time and storage. The most practical example involve fixed windows, i. e. , the window is always set up in the same place within the image. This usually means that some sort of fixturings must be used to identically position each workpiece so the consistency of the window system is maintained. Adaptive windowing can also be used. The system is able to select the appropriate window out of context. A search of the entire image detects known landmarks that identify the position and orientation of the subject workpiece. The landmarks can then be used by the system to find the window area of interest and proceed as in a fixed window situation. Whether the window is fixed or adaptive, the size can be adjusted for the purposes of the application. A single pixel can be used to pinpoint a colour or grey level with a single sample, though because of the possibility of spurious readings, it is more practical to sample a reasonable number of pixels and compute an average grey level.

窗口化 这是一种集中分析小视野图像的方法，能够节省计算机运行和存储时的资源。最实用的例子包括固定窗口，即总是把窗口设置在图像内的相同位置。这通常意味着必须使用某种夹具来使各工件的位置相同，以此来保持窗口系统的一致性。此外，也可以使用自适应窗口，使系统能够在上下范围之外选择适当的窗口。对整个图像进行搜索可检测出已知地标，这些地标可用于识别目标工件的位置和方向。然后系统可以使用这些地标来找到感兴趣的窗口区域，接着可以像固定窗口的情况中那样继续进行处理。无论窗口是固定的还是自适应的，都可以根据应用的目的调整尺寸。单个像素可通过单个样本来确定其颜色或灰度等级。由于存在虚假读数的可能性，因此通过采样恰当数量的像素并计算其平均灰度等级更为实际。

Thresholding A grey scale image can be further reduced by converting to a binary image where each pixel is represented by 1 or 0 (white or black). The threshold is the value that is used to change a grey scale

阈值化 通过将灰度图像转换为二值图像可以进一步将其减弱，其中二值图像的每个像素由 1 或 0（白色或黑色）表示。阈值是用于

image to a binary image. For example, if the threshold value is 150 then any values lower than this will become 0 and any value higher than this will become 1.

将灰度图像改为二值图像的值。例如，如果阈值为 150，那么任何小于此值的值将变为 0，任何大于此值的值将变为 1。

Fig. 8-14 shows the colour image, the grey scale image and the binary image of a part captured by the camera. It can be seen that the binary image is incomplete. This is due to glare (bright light) shining on the right hand side of the object, so the pixels there are brighter than the rest of the part and are being identified as background in the binary image. In order to improve the binary image, a different threshold value can be chosen. One way to do this is to construct a histogram of the pixels in the image. The histogram plots the frequency of each grey level and allows a considered choice of threshold value.

图 8-14 显示了由相机拍摄的某部件的彩色图像、灰度图像以及二值图像。从图中可以看出二值图像是不完整的，这是由于眩光（明亮的光线）照射在物体的右侧，导致那里的像素比其余部分更亮，从而使这部分在二值图像中被识别为背景。为了改善二值图像，可以选择不同的阈值。一种方法是构建图像中像素的直方图，其表示出每个灰度级的频率，并以此来选择阈值。

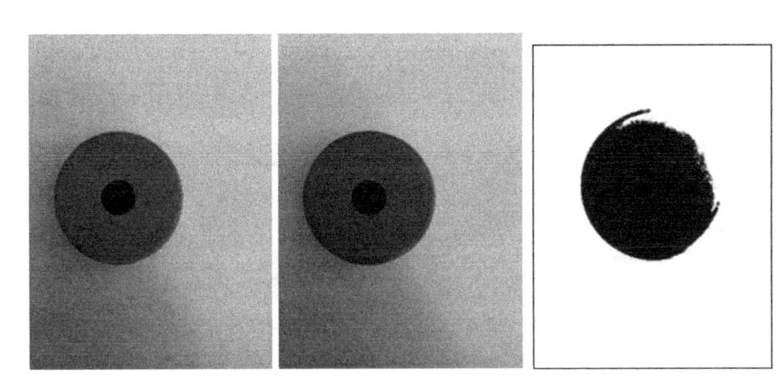

Fig. 8-14 Colour image, grey scale image and binary image

图 8-14 彩色图像、灰度图像和二值图像

In Fig. 8-15, the new threshold value chosen from the histogram (200) has resulted in a more complete image, but with some extra dark areas in the lower left hand corner.

在图 8-15 中，从直方图中选择新阈值（200）之后使图像变得更完整，但在左下角留下了一些多余的暗区。

Edge mapping It processes an image in order to identify the edge pixels and links them to produce a boundary map.

边缘映射 其处理图像以识别其边缘的像素并将它们联合起来以产生边界图。

- Edge information in an image is found by looking at the relationship between pixel and its neighbours.
- If a pixel's greylevel value is similar to those around

- 通过查看像素与其相邻像素之间的关系来找到图像中的边缘信息。
- 如果像素的灰度值与其周围像素

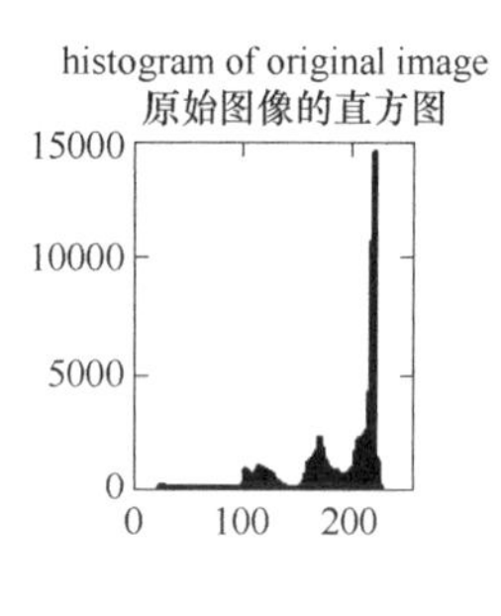

Fig. 8-15 Histogram and filtered image

图 8-15 直方图和滤波后的图像

it, there is probably not an edge at that point.

的值相似，则该点可能没有边缘。

• If a pixel has neighbours with widely varying grey levels, it may present an edge point.

• 如果像素的灰度级差异很大，可能会出现边缘点。

• Convolution mask is applied and gives discrete approximations of the rate of change of the grey scale values.

• 应用卷积掩膜，给出灰度值变化率的离散近似值。

Fig. 8-16 shows the result of edge mapping. A high pass filter is applied.

图 8-16 为部件边缘映射的结果，其中应用了高通滤波器。

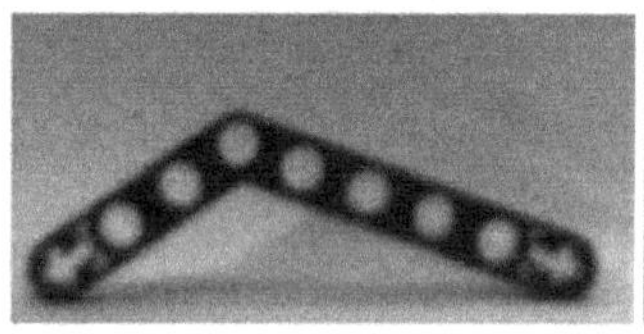

(a) original image
(a) 原始图像

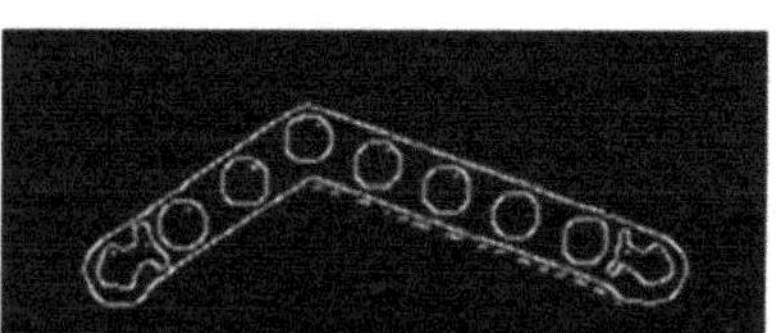

(b) processed image showing edge mapping
(b) 边缘映射处理后的图像

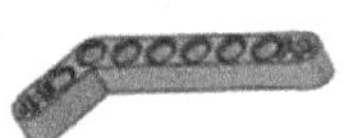

(c) component
(c) 部件

Fig. 8-16 The reasult of edge mapping

图 8-16 边缘映射结果

Pattern recognition

模式识别

① Template/pattern matching: the captured image is compared to computer stored model and the percentage match required is programmed to determine if a match is successfully found or not. This process is sensitive to object orientation, lighting, focus and size.

① 模板/模式匹配：将获取的图像与计算机存储的模型进行比较，并对最终所需的百分比匹配度进行编程以确定是否成功找到匹配项。此过程对物体的方向、采光、焦点和尺寸大小较为敏感。

② Feature weighting: several features combined into a single measure by assigning a weight to each feature according to its relative importance in identifying the object. The score is compared with the score of an ideal

② 特征加权：根据识别对象的相对重要性为对象的每个特征分配不同的权重，从而将多个特征组合成单个度量值。然后，将加权值

object as stored in the vision system memory.

与存储在视觉系统存储器中的理想对象的值进行比较。

8.4 Data Obtain and Process

8.4 数据获取及处理

Measurement system values can be outputted to the following:

测量系统的值可以输出到以下终端：

- PC with spreadsheet program for display and storage.
- PLC for data storage.
- PLC for automatically adjustment of the process.
- Process control charts.

- 带有用于显示和存储功能并具有电子表格程序的 PC。
- 用于数据存储的 PLC。
- 用于自动调整过程的 PLC。
- 过程控制图表。

【Example 8-1】 Adhesive dispense vision system sends adhesive location and thickness data to an Excel spread sheet which is configured to control real time data of the chart. When the adhesive thickness is monitored by the vision system and fed back to the PLC, the whole gluing process automatically adjusted to ensure process is in control.

【例 8-1】 黏合剂分配的视觉系统将黏合剂位置和厚度数据发送到 Excel 电子表格程序，从而控制图表的实时数据。当视觉系统监控到黏合剂厚度并反馈给 PLC 时，将会自动调整涂胶工艺以确保工艺在控制范围内。

8.5 Typical Vision System Algorithm Process

8.5 典型视觉系统算法过程

- The vision system sents the ready signal to production module.
- The production module moves component into position.
- The module sends signal to vision system to capture image.
- Values for each camera pixel are read and stored.
- Threshold pixel values.
- Image within search window is analyzed to see if pre-programmed pattern is found.
- If pattern is found, calculations are preformed and the data is sent to appropriate source.
- The system sends a signal to the production module to notify that its operation is completed.
- If pattern is not found, the vision system sends a message to the production module to notify.

- 视觉系统将准备好的信号发送到生产模块。
- 生产模块将组件移动到位。
- 模块将信号发送到视觉系统以获取图像。
- 读取并存储每个相机像素的值。
- 对像素值进行阈值化处理。
- 分析搜索窗口中的图像，看是否能找到预编程模式。
- 如果找到该模式，则执行计算并将数据发送到恰当的来源处。
- 系统向生产模块发送信号以通知其操作已完成。
- 如果未找到该模式，视觉系统会向生产模块发送消息通知。

8.6 Vision Limitation

- Two-dimensional measurement (used at times to monitor three-dimensional processes, e.g., adhesive dispensed).
- Material colour, surface finish and texture all have big effect on the system.
- System calibration may be complex.
- For high speed production equipment, the time to capture and process image may be too long.
- Expensive.
- Need trained personnel to maintain the system.

8.6 视觉局限性

- 二维测量（有时用于监控三维过程，例如黏合剂的分配）。
- 材料颜色，表面光洁度、纹理都对系统有很大影响。
- 系统校准可能很复杂。
- 对于高速生产设备，捕获时间和过程映像可能太长。
- 价格昂贵。
- 需要专门的培训人员维护系统。

8.7 Vision Calibration

- Calibration only necessary if the vision system is required to perform measurement on the component.
- Calibration is a process of teaching the vision system 'real world coordinates'. Essentially teaching the system that 1 pixel is equal to a given physical dimension.
- Use calibration part, part should be accurately fabricated, accurately measured; have similar material, colour, surface finish; have similar weight (to aid in fixturing repeatability).

8.7 视觉校准

- 仅在视觉系统需要对组件进行测量时才需要校准。
- 校准是教会视觉系统理解“真实世界坐标系”的过程。实质上相当于教系统1像素等于给定的物理尺寸。
- 使用校准部件，部件应精确制造，精确测量，具有相似的材料、颜色、表面光洁度，具有相似重量（有助于固定的重复性）。

8.8 Image Processing in Matlab

Some typical Matlab codes for image processing are presented as follows:

8.8 Matlab图像处理

一些典型的用于图像处理的Matlab代码如下所示：

```
Q=imread('mps Photo 1.jpg');
P=imresize(Q, 0.5);
figure, imshow(P), title('original image');
J=rgb2gray(P);
figure, imshow(J), title('grayscale image');
BW=im2bw(J);
figure, imshow(BW), title('BW image');
level=0.6;
BW1=im2bw(J, level);
```

```
figure, imshow (BW1), title ('threshold image');
h=ones (30, 30) /900;
S=imfilter (BW1, h);
S=~S;
figure, imshow (S), title ('filtered image');
Stats=regionprops (S, 'area', 'MajorAxisLength', 'EquivDiameter')
[pixelCount grayLevels] = imhist (J);
subplot (3, 3, 2);
bar (pixelCount); title ('Histogram of original image');
xlim ( [0 grayLevels (end) ] );
```

Common commands for image processing in Matlab are shown in Table 8-2.

Matlab中处理图像的常用命令见表8-2。

Table 8-2 Common commands in Matlab

表8-2 Matlab常用命令

imread	reads a particular image 读取特定图像
imresize	resizes an image 调整图像大小
rgb2gray	converts an image from RGB to grayscale 将图像从RGB转换为灰度
im2bw	converts an image to a binary image (black and white) 将图像转换为二值图像（黑白）
level	sets the value for the threshold 设置阈值的值
imfilter	applies an image to the filter 将滤波器应用于图像
regioprops	returns the properties of the image 返回图像的属性
imhist	plots a histogram 绘制直方图

Chapter 9 Numerical Control Technique
第9章　数控技术

Key words 重点词汇

CNC machine tool　数控机床	CNC milling　数控铣削
numerical control　数字控制	CNC interpolation　数控插补
standard coordinate system　标准坐标系	pulse equivalent　脉冲当量
machine coordinate system　机床坐标系	flexible manufacturing　柔性制造
linear interpolation　直线插补	integrated manufacturing　集成制造

9.1　Introduction of CNC Machine Tool

9.1.1　Definition of CNC Machine Tool

Numerical control (NC) is digital control. Numerical control technology is a kind of technology that controls one or more mechanical equipment with the control program formed by digital signals. Computer numerical control (CNC) is the NC that performs some or all of the basic digital control functions using a dedicated storable program computer.

CNC machine tool, in short, is the machine tool with NC technology. It means that all kinds of actions of machine tool, the shape and size of workpiece and other functions of machine tool are represented by some digital codes. These digital codes are input to the NC system through information carrier. After decoding, calculation and processing, NC system sends out corresponding action instructions, automatically controls the relative movement of cutter and workpiece of machine tool, so as to process the required workpiece. Therefore, CNC machine tool is a kind of

9.1　数控机床简介

9.1.1　数控机床的定义

数控（numerical control，NC）即数字控制，是用数字信号组成的控制程序对一台或多台机械设备进行控制的一门技术。计算机数字控制（CNC）则是指利用一个专用的、可存储程序的计算机执行一些或全部基本数字控制功能的数字控制。

数控机床，简单来说，就是采用了数控技术的机床，它把机床的各种动作，工件的形状、尺寸以及机床的其他功能用一些数字代码表示，把这些数字代码通过信息载体输入给数控系统，数控系统经过译码、运算以及处理，发出相应的动作指令，自动地控制机床上刀具与工件的相对运动，从而加工出所需要的工件。因此，数控机床是一种具有数控系统的

automatic machine tool with NC system, which belongs to the most typical mechatronic product. Fig. 9-1 and Fig. 9-2 are machining center and CNC lathe respectively.

自动化机床，是最典型的机电一体化产品。图 9-1 和图 9-2 分别为加工中心及数控车床。

Fig. 9-1 Machining center

图 9-1 加工中心

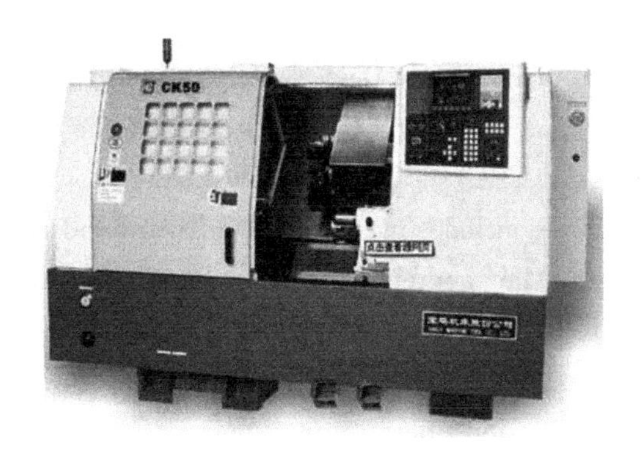

Fig. 9-2 CNC lathe

图 9-2 数控车床

9. 1. 2 Composition of CNC Machine Tool

9. 1. 2 数控机床的组成

9. 1. 2. 1 Components of CNC Machine Tool

9. 1. 2. 1 数控机床的组成部件

CNC machine tool is mainly composed of NC system, drive device, machine body, auxiliary device, etc.

数控机床主要由数控系统、驱动装置、机床本体、辅助装置等几个部分组成。

(1) NC System

(1) 数控系统

NC system (CNC device) is the control core of CNC machine tool, generally is a special computer. The basic work flow of NC system of machine tool is shown in Fig. 9-3. The NC system of machine tool is composed of machining instruction program, computer

数控系统（CNC 装置）是数控机床的控制核心，一般是一台专用的计算机。机床数控系统的基本工作流程如图 9-3 所示。机床数控系统是由加工指令程序、计算机控

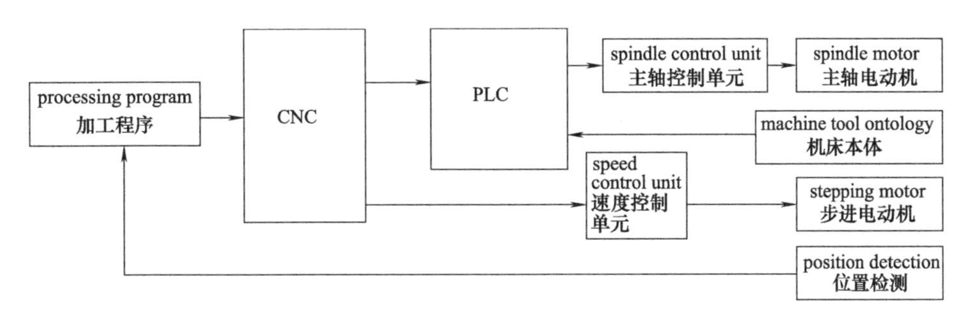

Fig. 9-3 Basic work flow of NC system of machine tool

图 9-3 机床数控系统基本工作流程

control device, programmable logic controller, spindle control unit, speed control unit and position detection device. And its core part is computer control device.

制装置、可编程逻辑控制器、主轴控制单元、速度控制单元及位置检测装置等组成，其核心部分是计算机控制装置。

Computer Control Device It consists of hardware and software. The main body of hardware is the computer, including CPU, I/O part and position control part. The software includes management software and control software. Management software includes input/output, display and diagnosis programs, control software includes decoding, tool compensation, speed control, interpolation operation and position control programs.

计算机控制装置 由硬件和软件两部分组成。硬件的主体是计算机，包括中央处理器、输入/输出部分和位置控制部分。软件有管理软件和控制软件。管理软件包括输入/输出、显示和诊断程序等；控制软件包括译码、刀具补偿、速度控制、插补运算和位置控制等程序。

Programmable Logic Controller It is abbreviated as PLC, which is an electronic system of digital operation, specially designed for application in industrial environment. PLC is between the numerical control device and the machine tool. It processes the I/O signals of the numerical control device and the machine tool to control and decoding the auxiliary function, spindle speed and tool function. That is to control the start, stop, turning and rotating speed of the spindle, the replacement of the cutting tool, the clamping and loosening of parts, hydraulic pressure, cooling, lubrication, pneumatic, etc., according to the predetermined logical sequence.

可编程逻辑控制器 简称PLC，是一种数字操作的电子系统，专门在工业环境下应用。PLC处于数控装置与机床之间，对数控装置和机床的输入/输出信号进行处理，实现辅助功能、主轴转速及刀具功能的控制和译码。即按照预先规定的逻辑顺序对启动、停止、转向、转速、刀具的更换、零件的夹紧松开、液压、冷却、润滑、气动等进行控制。

Speed Control Unit of the Spindle It mainly controls the rotation movement of the spindle of the machine tool.

主轴控制单元 主要控制机床主轴的旋转运动。

Speed Control Unit It is composed of the alternating current (AC) servo motor, speed detecting element and speed control element. The speed control unit mainly controls the cutting feed motion of each coordinate axis of the machine tool.

速度控制单元 由交流伺服电动机、速度检测元件和速度控制元件组成。速度控制单元主要控制机床各坐标轴的切削进给运动。

(2) Drive Device

(2) 驱动装置

The driving device is the driving part of the actuator of

驱动装置是数控机床执行机构的驱

CNC machine tool, including the spindle motor, feed servo motor, driver, etc.

动部分，包括主轴电动机、进给伺服电动机、驱动器等。

(3) Machine Body

(3) 机床本体

The main body is the mechanical part of CNC machine tool, mainly including supporting parts (bed, column), active parts (spindle box), feed moving parts (table slide plate, tool holder), etc.

机床本体是数控机床加工运动的机械部分，主要包括支承部件（床身、立柱）、主动部件（主轴箱）、进给运动部件（工作台滑板、刀架）等。

(4) Auxiliary Device

(4) 辅助装置

The auxilary device is the matching parts of CNC machine tool, including tool magazine, hydraulic device, pneumatic device, cooling system, chip removal device, clamp, tool change manipulator, etc.

辅助装置指数控机床的配套部件，包括刀库、液压装置、气动装置、冷却系统、排屑装置、夹具、换刀机械手等。

9.1.2.2 Basic functions of NC system

9.1.2.2 数控系统的基本功能

The NC system, namely the position control system, has three basic functions:

① Input function: it refers to the processing program of the part and the input of various parameters.

② Interpolation function: this is a method to determine some intermediate points between known points of the actual contour or track (such as line, arc or other curve) of the machined part.

③ Servo control: the position feed pulse or feed speed command sent by the computer is transformed and amplified into the stepping or the rotation of the AC servo motor, thus driving the table of the machine tool to move.

数控系统即位置控制系统，具有3个基本功能：

① 输入功能：指零件加工程序和各种参数的输入。

② 插补功能：在加工零件的实际轮廓或轨迹（如直线、圆弧或其他曲线）的已知点之间确定一些中间点的方法。

③ 伺服控制：将计算机送出的位置进给脉冲或进给速度指令变换和放大后转化为步进或交流伺服电动机的转动，从而带动机床工作台移动。

9.1.2.3 Working process of NC system

9.1.2.3 数控系统的工作过程

The NC system recognizes and decodes the processing program of the input parts, control parameter and compensation data, then performs the required logic operation, sends out the corresponding command pulse, controls the driving device of the machine tool, and operates the machine tool to achieve the expected processing function.

数控系统对输入的零件加工程序、控制参数、补偿数据等进行识别和译码，并执行所需要的逻辑运算，发出相应的指令脉冲，控制机床的驱动装置，操作机床实现预期的加工功能。

9.1.3 Working Principle of CNC Machine Tool

Fig. 9-4 shows the main process of CNC machine tool machining. During the machining processing of CNC machine tool, according to the requirements of part drawings and processing process, the requirements of the movement amount, speed, action sequence of cutting tools and parts of the machine tool, the speed and rotational direction of the spindle, and cooling behavior are compiled into a program list in the form of some specified CNC codes. Thus they are input into the special computer of the machine tool. Then, according to the input instructions, the NC system compiles, calculates and performs logical process. And various signal commands are output to control each part of the machine tool to perform prescribed displacements and sequential actions, and then process various parts of different shapes.

9.1.3 数控机床加工的工作原理

图 9-4 展示了数控机床加工的主要过程。数控机床加工时，根据零件图要求及加工工艺过程，将所用刀具及机床各部件的移动量、速度、动作先后顺序、主轴转速、主轴旋转方向及冷却要求，以规定的数控代码形式编制成程序单，并输入到机床专用计算机中。然后，数控系统根据输入的指令，进行编译、运算和逻辑处理，输出各种信号指令，控制机床各部件执行规定的位移和有顺序的动作，加工出各种不同形状的零件。

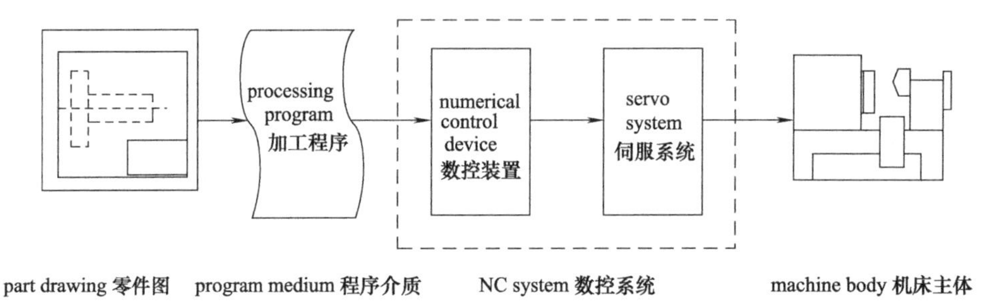

Fig. 9-4 The basic principle of CNC

图 9-4 数控机床的基本原理

9.1.4 Features of CNC Machining

CNC machine tool is a kind of machine tool with high automation, high efficiency and high precision. It performs automatic processing in accordance with the CNC processing program of the processed parts. When the production object changes, only the CNC processing program made to be changed. The production preparation cycle is short and it is the preferred flexible equipment for modern production systems. The CNC

9.1.4 数控加工的特点

数控机床是一种高度自动化、高效率、高精度的机床。它按照被加工零件的数控加工程序进行自动加工。当生产对象变化时，只需要改变数控加工程序，生产准备周期短，是现代生产系统中柔性设备的首选。数控机床使用工装夹具少，减少了加工过程中的工

machine tool occupies less fixture and can reduce the work-in-process, which is beneficial to improving the economic benefits of the enterprise.

作量，有利于提高企业的经济效益。

CNC machine tool has good consistency, high machining accuracy, and stable machining quality when machining parts. The CNC machine tool automatically processes the parts according to the predetermined processing program. The processing process does not require manual intervention. CNC machine tool has good rigidity and high precision. And it can also use software to correct and compensate for the precision. Therefore, it can obtain higher machining precision and repeated precision than the machine tool itself.

数控机床加工零件的一致性好，加工精度高，加工质量稳定。数控机床按预定的零件加工程序自动加工，加工过程不需要人工干预。数控机床本身的刚度好、精度高，而且还可利用软件进行精度校正和补偿。因此可以获得比机床本身精度还要高的加工精度和重复精度。

CNC machine tool has a strong ability to process complex surfaces. It plays an extremely important role in the precision machining of complex curved parts in industries such as aviation, aerospace, mould, worm blade and propeller blade.

数控机床具备很强的复杂曲面加工能力。在航空、航天、模具、蜗轮叶片和螺旋桨叶片等行业的复杂曲面零件精密加工中发挥极其重要的作用。

The production efficiency of CNC machine tool is high. CNC machine tool can use large cutting parameters, save the mobile time, and realize the functions of automatic speed change, automatic tool change, automatic workpiece exchange and other auxiliary operation automation, which can shorten the auxiliary time. Therefore, its productivity is 3～4 times higher than the ordinary machine tools, even higher, and the labor intensity of workers is greatly reduced.

数控机床的生产效率高。数控机床可以采用大切削用量，节省机动时间，能够实现自动变速、自动换刀、自动交换工件和其他辅助操作自动化等功能，使辅助时间缩短。故其生产率比普通机床高3～4倍，甚至更高，且大大减轻了工人的劳动强度。

The CNC machine tool realizes the process compounding. After equipped with CNC turntable, the compound CNC machine tool has a variety of technological capabilities such as turning, milling, boring, drilling, grinding, etc. It realizes "one-time clamping, full-sequence processing". It reduces the installation error, saves auxiliary time of transportation, measurement, and installation of process flow and save the area occupied by the machine tool.

数控机床实现了工序复合化。复合化数控机床配备数控转台后，具备车、铣、镗、钻、磨等多种工艺能力，实现"一次装夹、全序加工"，减少装夹误差，节约工序流转的运输、测量、装夹等辅助时间，节省机床占地面积。

9. 1. 5 Classification of CNC Machine Tool

9. 1. 5 数控机床分类

There are many kinds of CNC machine tools, such as turning, milling, drilling and reaming, grinding, wire cutting, machining center, etc. (Fig. 9-5). There are also many classification methods, and some of them are provided as follows.

数控机床种类很多，如车削类、铣削类、钻铰类、磨削类、线切割、加工中心等（图 9-5），其分类方法也很多，大致有以下几种。

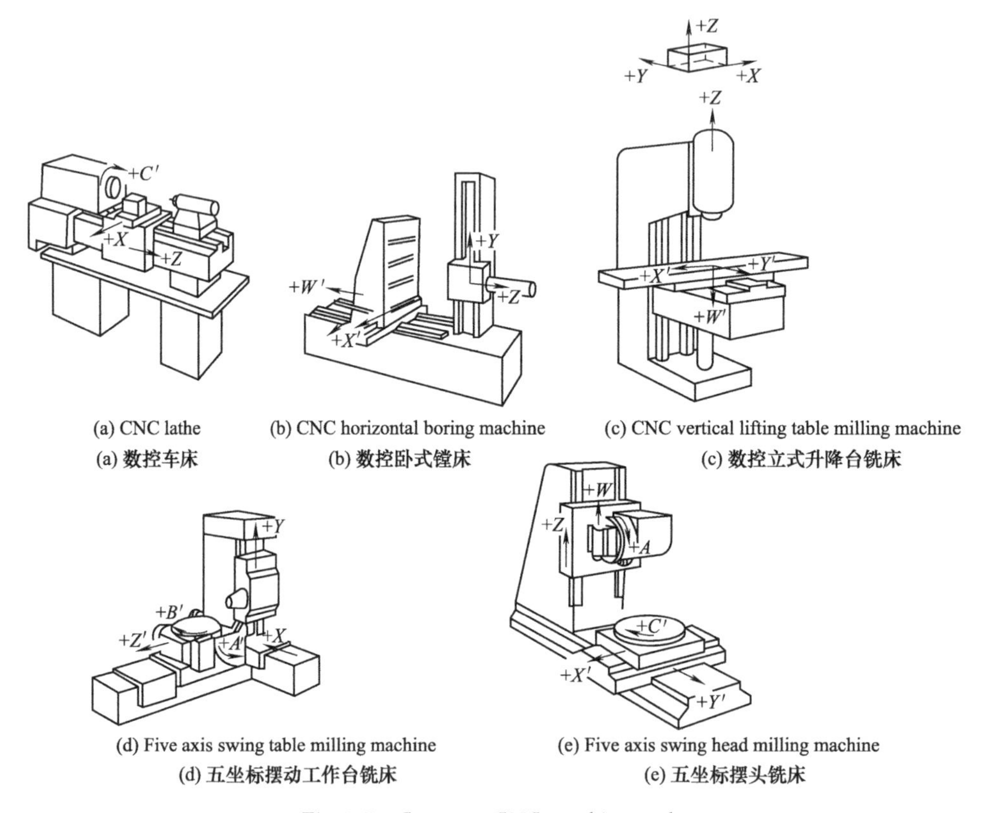

Fig. 9-5 Common CNC machine tools

图 9-5 常用数控机床

(1) Classification According to Relative Motion trajectory of Control Tool and Part

(1) 按控制刀具与零件相对运动轨迹分类

• CNC machine tool with point control or position control: it can only control the worktable or tool to move from one position to another accurately. During the movement, no processing is performed. Each motion axis can move at the same time or in sequence, as shown in Fig. 9-6 (a). Such as CNC boring, drilling, punching, spot welding machine and bending machine are all this kind of machine.

• 点位控制或位置控制数控机床：它只能控制工作台或刀具从一个位置精确地移动到另一位置，在移动过程中不进行加工，各个运动轴可以同时移动，也可以依次移动，如图 9-6(a) 所示。数控镗、钻、冲机床，数控点焊机及数控折弯机等均属此类。

• CNC machine tool with contour control: it can continuously control two or more coordinate axes at the same time, and has interpolation function. The worktable or cutter can move and process at the same time, as shown in Fig. 9-6 (b) and (c). For example, CNC milling, turning, grinding and machining center are typical CNC with contour control. The CNC flame cutting machine, CNC wire cutting and CNC plotter also use contour control system.

•轮廓控制数控机床：它能够同时对两个或两个以上坐标轴进行连续控制，具有插补功能，工作台或刀具边移动边加工，如图 9-6 (b)、(c) 所示。数控铣床、车床、磨床及加工中心等是典型的轮廓控制数控机床，数控火焰切割机、数控线切割及数控绘图机等也都采用轮廓控制系统。

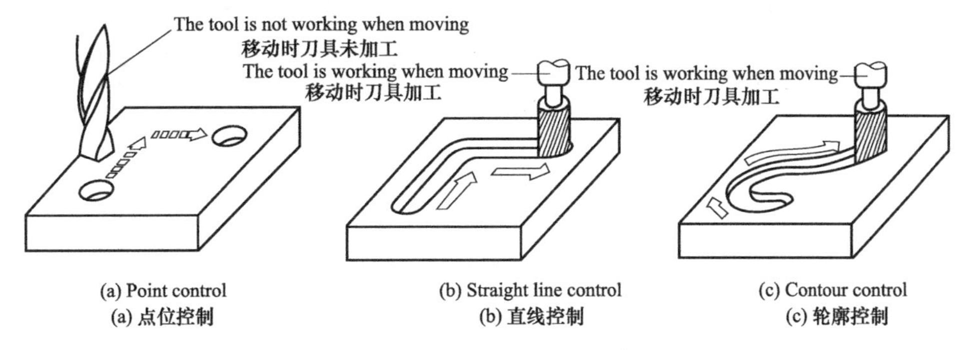

Fig. 9-6 Trajectory control

图 9-6 运动轨迹控制

(2) Classification by Processing Method

(2) 按加工方式分类

• Metal cutting: such as CNC turning, drilling, boring, milling, grinding, machining center, etc.

• Metal forming category: such as CNC bending machine, bending machine, four-turn pressure machine, etc.

• Special processing: such as CNC wire cutting maching, electric spark cutting machine, laser cutting machine, etc.

• Other categories: such as CNC flame cutting machine, three-coordinate measuring machine, etc.

•金属切削类：如数控车床、钻床、镗床、铣床、磨床、加工中心等。

•金属成形类：如数控折弯机、弯管机、四转头压力机等。

•特殊加工类：如数控线切割机、电火花机、激光切割机等。

•其他类：如数控火焰切割机、三坐标测量机等。

(3) Classification by Number of Control Axes

(3) 按控制坐标轴数分类

• Linkage of two axes: mostly used for processing various curve profiles, such as CNC lathe.

• Linkage of three axes: mostly used for machining curved surface parts, such as CNC milling machine, CNC grinding machine.

• Linkage of multiple axes: linkage of four or five axes, mostly used for machining parts with complex

•两轴联动：多用于加工各种曲线轮廓，如数控车床。

•三轴联动：多用于加工曲面零件，如数控铣床、数控磨床。

•多轴联动：四轴或五轴联动，多用于加工形状复杂的零件。图

shapes. Fig. 9-7 shows two different types of five axis CNC machine tool.

9-7 所示为两种不同类型的五轴联动数控机床。

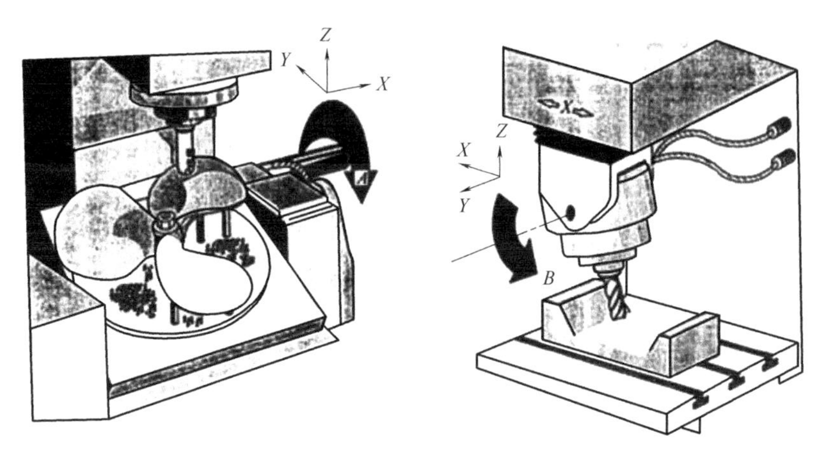

Fig. 9-7 Five axis CNC machine tool

图 9-7 五轴联动数控机床

(4) Classification via the Control Mode of Driving System

(4) 按驱动系统的控制方式分类

- Open-loop control CNC machine tool: the working principle of open-loop control CNC is shown in Fig. 9-8.

- 开环控制数控机床：开环控制数控机床的工作原理如图 9-8 所示。

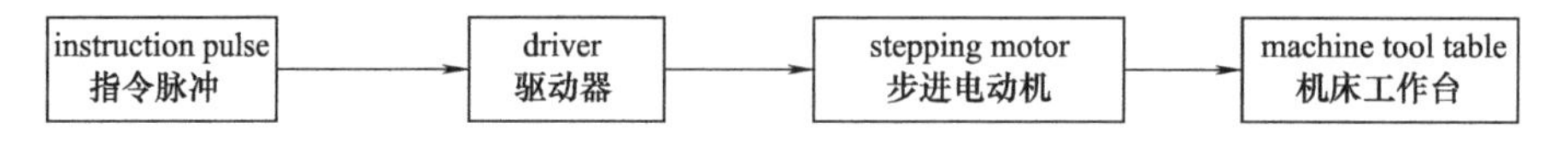

Fig. 9-8 The working principle open-loop control CNC

图 9-8 开环控制数控机床工作原理

The open-loop control CNC machine tool does not have a position detection feedback device. Generally, the power stepping motor is used as the actuator. The pulse output by the CNC device passes through the ring distributor and the drive circuit to make the stepping motor turn around the corresponding step angle, then the screw is driven to rotate through the reduction gear. Finally the ratation angle is converted into a linear displacement of the moving part. It has quick reaction, convenient debugging, relatively stable and simple maintenance. However, the system has no compensation and correction for the error of moving parts. The step error of stepping motor, the transmission error of gear and lead screw will be reflected in the accuracy of the processed parts, so the accuracy is relatively low. These kinds of CNC machine tools are mostly economic.

开环控制数控机床不带位置检测反馈装置，通常使用功率步进电动机作为执行机构。数控装置输出的脉冲通过环形分配器和驱动电路，使步进电动机转过相应的步距角，再经过减速齿轮，带动丝杠旋转，最后转换为移动部件的直线位移。其反应快，调试方便，过程稳定，维修简单。但系统对移动部件的误差没有补偿和校正，步进电机的步距误差、齿轮与丝杠等的传动误差都将反映到被加工零件的精度中，所以精度比较低。此类数控机床多为经济类机床。

• Closed-loop control CNC machine tool: the closed-loop control CNC machine tool is equipped with a detection feedback device. And the position detector is installed on the moving parts of the machine tool. During the processing, the actual running position detected is fed back to the CNC device. Compared with the input command position, the different value is used to control the moving parts, and thus its precision is high. Theoretically, the control accuracy of the closed-loop system mainly depends on the accuracy of the detection device. But this does not mean that it can reduce the requirements of the structure of the machine tool and the transmission chain, the insufficient rigidity and clearance of the transmission system, the creeping of the guide rail and other factors will increase the difficulty of debugging. The quality of the closed-loop control system will decline or even cause oscillation when it is severe. Therefore, the design and adjustment of the closed-loop system are very difficult. Such machine tools are mainly used in some boring and milling machines, lathes and machining centers with high precision requirements. The working principle of closed-loop control of CNC machine tool is shown in Fig. 9-9.

• 闭环控制数控机床：闭环控制数控机床带有检测反馈装置，位置检测器安装在机床的运动部件上，它将加工中检测到的实际运行位置反馈到数控装置中，与输入的指令位置相比较，用差值对移动部件进行控制，其精度高。从理论上说，闭环系统的控制精度主要取决于检测装置的精度，但这并不意味着可以降低机床的结构与传动的要求。传动系统的刚性不足及间隙、导轨的爬行等因素将增加调试的困难，严重时会使闭环控制系统的品质下降甚至引起振动。故闭环系统的设计和调整都有较大的难度，此类机床主要用于一些精度要求较高的镗铣床、车床和加工中心等。闭环控制数控机床的工作原理如图9-9所示。

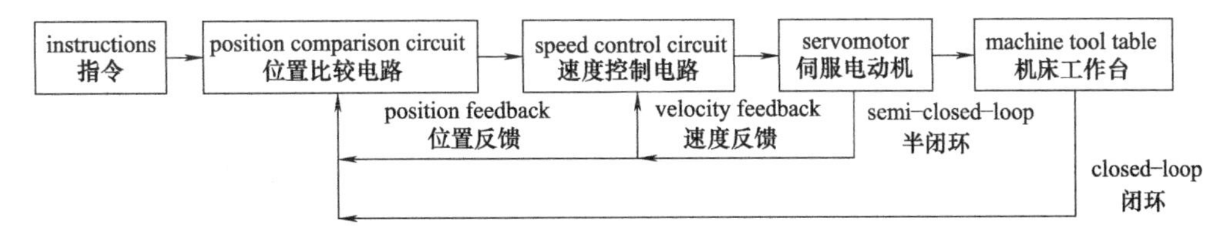

Fig. 9-9 The working principle of closed-loop and semi-closed-loop control CNC

图 9-9 闭环与半闭环控制数控机床的工作原理

• Semi-closed-loop control of CNC machine tool: the difference between semi-closed-loop control and closed-loop control is that the position detecting element is usually connected with the shaft of servo motor. This system does not directly measure the displacement of the worktable. But it indirectly measures the displacement of the worktable by detecting the rotation angle of the servo motor axis, and then feeds back to the numerical control device. Obviously, the

• 半闭环控制数控机床：半闭环控制数控机床与闭环控制不同的是位置检测元件通常与伺服电机轴固连。该系统不是直接测量工作台的位移量，而是通过检测伺服电机轴的转角，间接地测量工作台的位移量，然后反馈给数控装置。显然，半闭环控制系统的实际控制量是电机的转动，不包括

actual control quantity of the semi-closed-loop control system is the rotation of the motor, excluding the transmission parts such as reducer, ball screw and worktable. The accuracy of this part is guaranteed by the mechanical transmission accuracy of the feed system. Its characteristics are relatively stable and convenient debugging. And its accuracy is between open-loop and closed-loop. Therefore, it is widely used.

减速器、滚珠丝杠及工作台等传动部分，这一部分的精度由进给系统的机械传动精度来保证。半闭环控制数控机床的特点是比较稳定，调试方便，精度介于开闭环之间，现在被广泛地使用。

9.2 Fundamentals of CNC Machine Tool Programming

9.2 数控机床编程基础

9.2.1 Basic Concept of CNC Programming

9.2.1 数控编程的基本概念

A CNC machine tool automatically processes the parts according to the pre-programmed processing program. Therefore, CNC programming is to write the processing program sheet of the processing route and process parameters of the parts, the motion trajectory, the displacement, the cutting parameters (spindle speed, feed, cutting depth, etc.) and auxiliary functions (tool change, spindle forward and reverse, cutting fluid on and off, etc.) according to the command code and program format specified by the CNC machine tool. Thus they are input into the CNC device in order to control the machining parts.

数控机床按照事先编制好的加工程序，自动对零件进行加工。因此，数控编程就是把零件的加工工艺路线、工艺参数，刀具的运动轨迹、位移量、切削参数（主轴转速、进给量、切削深度等）以及辅助功能（换刀，主轴正反转，切削液开、关等）按照数控机床规定的指令代码及程序格式编写成加工程序单，并将其输入到数控机床的数控装置中，从而控制机床。

9.2.2 Steps of CNC Programming

9.2.2 数控程序编制的步骤

CNC programming refers to the whole working process from part drawing to obtaining CNC machining program. Its steps are shown in Fig. 9-10.

数控编程是指从零件图样到获得数控加工程序的全部工作过程，其步骤如图 9-10 所示。

(1) Analyze Part Drawing and Make Process Plan

(1) 分析零件图样和制定工艺方案

By analyzing the material, shape, size and technical requirements of the parts, the proper CNC can be selected to determine the processing sequence, processing route, clamping method, cutting tools, cutting parameters, etc.

对零件材料、形状、尺寸、技术要求等进行分析，选择合适的数控机床，确定加工顺序、加工路线、装夹方式、刀具、切削用量等。

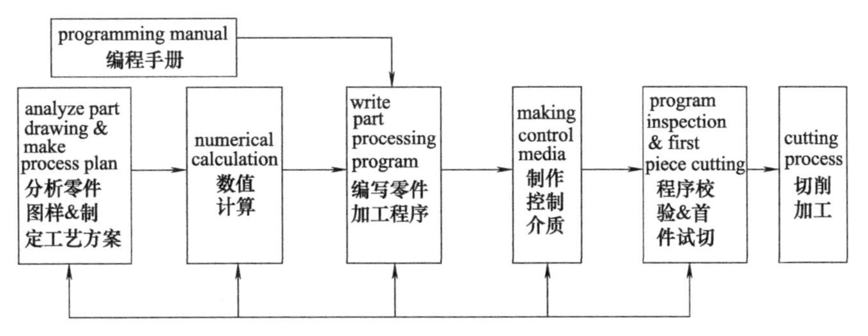

Fig. 9-10 Manual CNC programming steps

图 9-10 手动数控编程步骤

(2) Numerical Calculation

According to the determined processing route and processing error, the input data needed by the CNC can be calculated. The complexity of numerical calculation depends on tha of parts and the function of NC system. For a simple contour composed of lines and arcs, only the intersection or tangent point, start point, end point of geometric elements and the center coordinates of arcs need to be calculated, which can be done manually. For parts with complex shapes, such as spline curves, it is usually necessary to use computer and CAD/CAM software to calculate.

(2) 数值计算

根据已确定的加工路线和加工误差，计算出数控机床所需的输入数据。数值计算的复杂程度取决于零件的复杂程度和数控系统的功能。对于由直线和圆弧组成的简单轮廓，只需计算出几何元素的交点或切点、起点、终点和圆弧的圆心坐标等，这可由人工来完成。对于形状较复杂的零件，如样条曲线等，通常需要借助计算机和CAD/CAM软件来进行计算。

(3) Writing Part Processing Program

Based on the process analysis scheme, numerical calculation results and auxiliary operation requirements, the machining program can be compiled in accordance with the program instructions and formats specified by the NC system.

(3) 编写零件加工程序

根据工艺分析方案、数值计算结果以及辅助操作要求，按照数控系统规定的程序指令及格式编写出加工程序。

(4) Making Control Media

Making the control medium is to record the written program on the control medium, and input the CNC machining program on the control medium to the CNC through the input device of the machine tool (such as network port, USB port, etc.).

(4) 制作控制介质

制作控制介质就是将编写好的程序记录在控制介质上，并通过机床的输入装置（如网口、USB口等），将控制介质上的数控加工程序输入到数控机床中。

(5) Program Inspection and First Piece trial Cutting

(5) 程序校验与首件试切

In order to ensure the correctness of parts processing, the NC program must be verified and used to trial cutting before it can be used for formal processing. Generally, we can check the machining program by means of idle running and simulation machining, but these methods can not check the accuracy of the machined parts. To check the machining accuracy of the parts to be processed, usually through the first piece trial cutting, if the machining accuracy fails to meet the requirements, it is necessary to analyze the causes of the errors and take measures to correct them.

为了保证零件加工的正确性，数控程序必须经过校验和试切才能用于正式加工。通常可以采用机床空运行和模拟加工的方法来检查加工程序，但这些方法不能检验被加工零件的精度。为了检验被加工零件的加工精度，通常进行首件试切，若发现加工精度达不到要求，应分析其误差产生原因，采取措施加以纠正。

9.2.3 The Method of CNC Programming

9.2.3 数控程序编制的方法

There are two main methods of CNC programming: manual programming and automatic programming.

数控加工程序的编制方法主要有两种：手工编制程序和自动编制程序。

(1) Manual Programming

(1) 手工编程

Manual programming mainly refers to complete all stages of CNC programming work manually. The analysis of part drawings, the formulation of process routes, the selection of process parameters, the numerical calculation and the preparation of processing program sheets are all done manually.

手工编程是指主要由人工来完成数控编程中各个阶段工作的程序编制方法。分析零件图样、制定工艺路线、选用工艺参数、进行数值计算、编写加工程序单等都由人工来完成。

Manual programming requires the programmer not only to be familiar with the CNC instructions and programming rules of the CNC, but also to have certain CNC machining process knowledge and numerical calculation ability. Generally speaking, for parts with simple shape, the calculation amount is small, the program is short, and the manual programming is fast, simple and economical. Therefore, manual programming is widely used in point machining or plane contour composed of straight lines and arcs.

手工编程不仅要求编程人员熟悉所用数控机床的数控指令及编程规则，还要求其具备一定的数控加工工艺知识和数值计算能力。一般而言，对于形状简单的零件，计算量小、程序短，用手工编程快捷、简便、经济。因而手工编程广泛用于点位加工或加工由直线与圆弧组成的平面轮廓。

(2) Automatic Programming

(2) 自动编程

Automatic programming refers to the process that in the programming process, in addition to analyzing the part drawing and making the process plan manually, other CNC machining programs are compiled with computers and corresponding programming software. The commonly used CAD/CAM software for automatic programming includes Creo, UG, Cimatron, CAXA, Solidworks, etc.

自动编程是指在编程过程中，除了分析零件图样和制定工艺方案由人工进行外，其他数控加工程序均用计算机及相应编程软件编制的程序编制方法。自动编程常用的 CAD/CAM 软件有 Creo、UG、Cimatron、CAXA、SolidWorks 等。

When programming automatically, the programmer first uses CAD software for three-dimensional modeling, then uses CNC module of CAM software to input processing requirements such as process parameters. And then CAM to is employed generate tool path and post-processing program is utilized to generate processing program corresponding to specific CNC machine tools. Finally, the CNC processing program is transferred to CNC tools through DNC transmission software to realize online processing. Compared with manual programming, automatic programming has the advantages of short programming time, low labor intensity, low error probability and high programming efficiency. Therefore, it is suitable for the programming of parts with complex shape or composed of spatial surfaces.

自动编程时，编程人员先利用 CAD 软件进行三维造型，再利用 CAM 软件的 CNC 模块输入工艺参数等加工要求，然后由 CAM 产生刀具路径，进而再利用后置处理程序产生与具体数控机床对应的加工程序，最后利用 DNC 传输软件将数控加工程序传给数控机床，实现联机加工。与手工编程相比，自动编程具有编程时间短、编程人员劳动强度低、出错概率小、编程效率高等优点。因此，它适用于形状复杂或由空间曲面组成的零件的加工编程。

9.2.4 CNC Machine Tool Coordinate System

9.2.4 数控机床坐标系

In order to describe the movement of the machine tool and simplify the programming, the coordinate system and the movement direction of the machine tool have been standardized.

在数控编程时，为了统一描述机床的运动，简化程序编制，对数控机床的坐标系和运动方向进行了标准化。

9.2.4.1 Standard Coordinate System

9.2.4.1 标准坐标系

The standard coordinate system is determined as follows.

标准坐标系的确定原则如下。

(1) Provisions for Relative Movement of Machine Tools

(1) 机床相对运动的规定

No matter whether the tool or the part is moving when the machine tool is being processed, it is assumed that the part is stationary and the tool is relatively moving.

无论机床加工时刀具运动还是零件运动，都统一假设零件静止，刀具作相对运动。

(2) The Regulation of the Coordinate System of the Machine Tool

(2) 机床坐标系的规定

In order to determine the displacement and direction of forming motion and auxiliary motion on CNC machine tool, Cartesian coordinate system is usually used to describe the geometric space of machine tools, that is, machine coordinate system. The relationship between the X, Y and Z coordinate axes in the coordinate system is determined by the right-hand Cartesian rectangular coordinate system, as shown in Fig. 9-11. The three rotation coordinates of A, B and C are shown in Fig. 9-12, which are determined by the right-hand spiral rule.

为确定数控机床上成形运动和辅助运动的位移和方向，通常采用笛卡尔直角坐标系对机床几何空间进行描述，即机床坐标系。机床坐标系中 X、Y、Z 坐标轴的相互关系用右手笛卡尔直角坐标系决定，如图 9-11 所示。A、B、C 三个旋转坐标如图 9-12 所示，用右手螺旋法则确定。

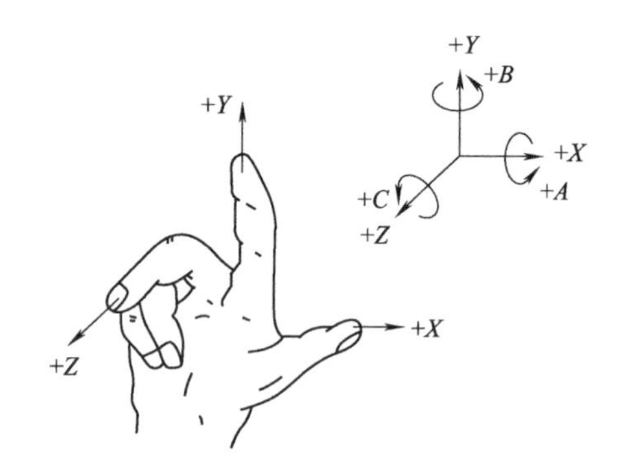

Fig. 9-11 Cartesian right-hand coordinate system
图 9-11 右手笛卡尔直角坐标系

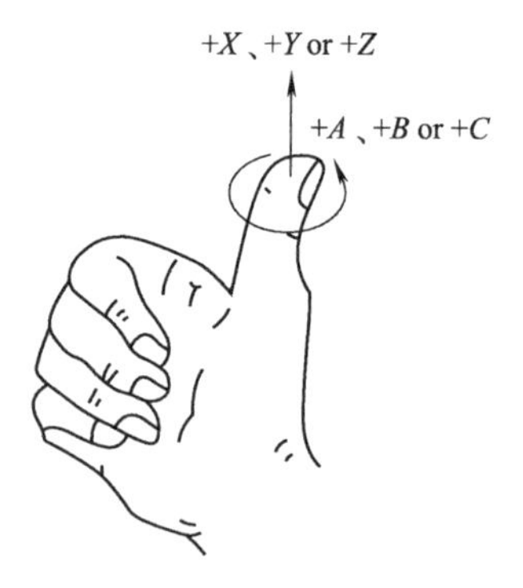

Fig. 9-12 Right-hand rule
图 9-12 右手螺旋法则

(3) Regulation of Movement Direction

(3) 运动方向的规定

Specify the direction of the tool away from the part as the positive direction of each coordinate axis. There are three moving coordinate axes on the CNC lathe shown in Fig. 9-13. The CNC milling machine shown in Fig. 9-14 has three motion axes.

规定刀具远离零件的方向为各坐标轴的正方向。图 9-13 所示的数控车床上有三个运动坐标轴。图 9-14 所示的数控铣床有三个运动坐标轴。

Fig. 9-13 Coordinate system of horizontal lathe
图 9-13 卧式车床的坐标系

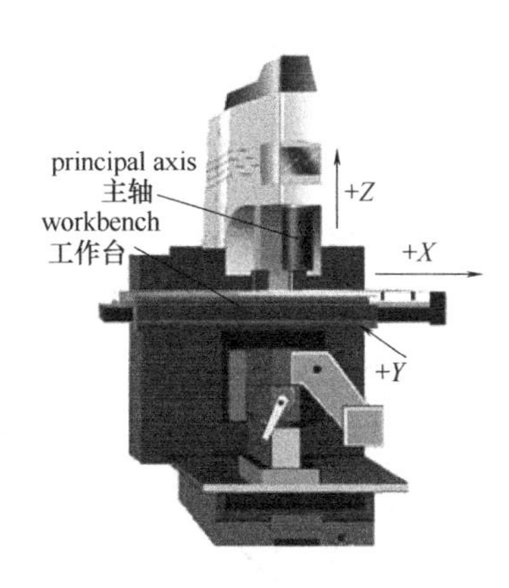

Fig. 9-14 Coordinate system of CNC vertical milling machine
图 9-14 数控立式铣床的坐标系

9. 2. 4. 2 Determination of Direction of Coordinate Axis

9. 2. 4. 2 坐标轴方向的确定

(1) Determine the Z-Axis

(1) 确定 Z 轴

The movement direction of Z-axis is determined by the spindle which transmits the cutting power, that is, the coordinate axis parallel to the spindle axis is Z-axis, and the positive direction of Z-axis is the direction in which the tool leaves the part.

Z 轴的运动方向是由传递切削动力的主轴决定的，即平行于主轴轴线的坐标轴即为 Z 轴，Z 轴的正向为刀具离开零件的方向。

(2) Determine the X-Axis

(2) 确定 X 轴

The X-axis is parallel to the clamping plane of the part, generally in the horizontal plane. There are two situations to be consider when determining the direction of the X-axis:

• If the part rotates, the direction of the tool leaving the part is the positive direction of X-axis, as shown in Fig. 9-15 (a).

• If the tool rotates, it can be divided into two situations: when the Z-axis is vertical, the observer faces the tool spindle to the column, and the +X movement direction points to the right, as shown in Fig. 9-15 (b), when the Z-axis is horizontal, the observer looks to the part along the tool spindle, and the +X

X 轴平行于零件的装夹平面，一般在水平面内。确定 X 轴的方向时，要考虑两种情况：

• 如果零件作旋转运动，则刀具离开零件的方向为 X 轴的正方向，如图 9-15(a) 所示。

• 如果刀具作旋转运动，则分为两种情况：Z 轴垂直时，观察者面对刀具主轴向立柱看，+X 运动方向指向右方，如图 9-15(b) 所示；Z 轴水平时，观察者沿刀具主轴向零件看，+X 运动方向

movement direction points to the left, as shown in Fig. 9-15 (c).

指向左方，如图 9-15(c) 所示。

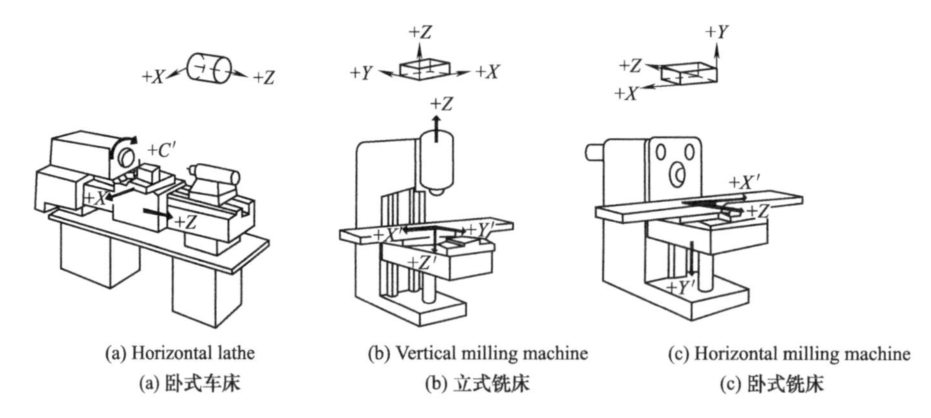

(a) Horizontal lathe (b) Vertical milling machine (c) Horizontal milling machine

(a) 卧式车床 (b) 立式铣床 (c) 卧式铣床

Fig. 9-15 Coordinate system of typical CNC

图 9-15 典型数控机床的坐标系

(3) Determine the Y-Axis

(3) 确定 Y 轴

After determining the positive direction of X-axis and Z-axis, the direction of Y-axis is directly determined according to the right-hand rectangular coordinate system.

在确定 X、Z 轴的正方向后，按照右手直角坐标系直接确定 Y 轴的方向。

(4) Determine the rotary axes A, B and C

(4) 确定回转轴 A、B、C

According to the determined X, Y and Z axes, use the right-hand screw rule to determine the three-axis coordinates of the rotation axes A, B and C.

根据已确定的 X、Y、Z 轴，用右手螺旋法则确定回转轴 A、B、C 三轴坐标。

9.2.5 NC Program Format

9.2.5 数控加工程序格式

(1) Program Start and End Characters

(1) 程序开始符、结束符

The program start and end characters are the same. In ISO code, they are % and in EIA code, they are EP. When writing, they should be listed separately.

程序开始符与结束符相同，ISO 代码中是%，EIA 代码中是 EP，书写时要单列一段。

(2) Program Name

(2) 程序名

There are two forms of program name: one is composed of the English letter O and 1～4 positive integers. The other is composed of a mixture of letters and numbers starting with English letters. The program name is gener-

程序名有两种形式：一种是由英文字母 O 和 1～4 位正整数组成；另一种是由英文字母开头，字母数字混合组成。程序名一般要求单

ally required to be listed in a single paragraph.

列一段。

(3) Procedural Subject

(3) 程序主体

The program subject is composed of several program segments. Each segment typically takes up one line.

程序主体是由若干个程序段组成的。每个程序段一般占一行。

(4) Program End Instruction

(4) 程序结束指令

The program terminator can be expressed by M02 or M30, and it is generally required to be listed in a single section.

程序结束指令可以用 M02 或 M30 表示，一般要求单列一段。

A general format example for the processing program:

加工程序的一般格式示例如下：

```
%                                   //start symbol 开始符
O1000                               //program name 程序名
N10 G00 G54 X50 Y30 M03 S3000       //procedural subject 程序主体
N20 G01 X88. 1 Y30. 2 F500 T02 M08
N30 X90
...
N300 M30                            //terminator 结束符
```

9.2.6 Program Segment Function Word

9.2.6 程序段功能字

Now we usually use the variable program segment format of word address. An example of the program segment format is shown below:

现在一般使用字地址的可变段格式。程序段格式举例如下：

N30 G01 X88. 1 Y30. 2 F500 S3000 T02 M08

(1) Sequence Number Word N

(1) 顺序号字 N

The sequence number is at the beginning of the program segment and consists of N and subsequent numbers. The function of sequence number:

- Check and retrieve the program.
- It can be used as the target of conditional turning, that is, as the name of the program segment with turning target.

顺序号位于程序段之首，由 N 和后续数字组成。顺序号的作用：

- 对程序进行校对和检索修改。
- 作为条件转向的目标，即作为转向目的程序段的名称。

(2) Preparatory Function Word G

The address character of the preparatory function word is G, also known as G function or G instruction, which is an instruction used to establish the working mode of the machine tool or control system. The following numbers are generally 1～3 positive integers, as shown in table 9-1.

(3) Dimension Word X \ Y \ Z

Dimension word X \ Y \ Z is used to determine the coordinate position of the tool movement end points of the machine tool.

(4) Feed Function Word F

The address character of the feed function word is F, it's used to specify the feed rate for the cut.

(5) Spindle Speed Function Word S

The address character of the spindle speed function word is S, which is used to specify the spindle speed, and the unit is r/min.

(6) Tool Function Word T

The address character of the tool function word is T which is employed to specify the number of the tool used in machining.

(7) Auxiliary Function Word M

The address character of auxiliary function word is M which is used to specify the switch action of auxiliary device of CNC machine tool, as shown in table 9-2.

(2) 准备功能字 G

准备功能字的地址符是 G，又称 G 功能或 G 指令，是用于建立机床或控制系统工作方式的一种指令。后续数字一般为 1～3 位正整数，如表 9-1 所示。

(3) 尺寸字 X \ Y \ Z

尺寸字 X \ Y \ Z 用于确定机床上刀具运动终点的坐标位置。

(4) 进给功能字 F

进给功能字的地址符是 F，用于指定切削的进给速度。

(5) 主轴转速功能字 S

主轴转速功能字的地址符是 S，用于指定主轴转速，单位为 r/min。

(6) 刀具功能字 T

刀具功能字的地址符是 T，用于指定加工时所用刀具的编号。

(7) 辅助功能字 M

辅助功能字的地址符是 M，用于指定数控机床辅助装置的开关动作，如表 9-2 所示。

Table 9-1 Common G instruction codes of preparation function

表 9-1 常用准备功能 G 指令代码

Code 代码	Group 分组	Function 功能	Code 代码	Group 分组	Function 功能
* G00	01	Point location 点定位	G55	14	Select No. 2 workpiece coordinate system 选用 2 号工件坐标系
* G01	01	Linear interpolation 直线插补	G56	14	Select No. 3 workpiece coordinate system 选用 3 号工件坐标系
G02	01	Clockwise arc interpolation 顺时针圆弧插补	G57	14	Select No. 4 workpiece coordinate system 选用 4 号工件坐标系
G03	01	Anticlockwise arc interpolation 逆时针圆弧插补	G58	14	Select No. 5 workpiece coordinate system 选用 5 号工件坐标系
G04	00	Suspend 暂停	G59	14	Select No. 6 workpiece coordinate system 选用 6 号工件坐标系
G09	00	Precise stop 精确停止	* G64	15	Cutting mode 切削方式
* G17	02	*XY* plane selection *XY* 平面选择	G65	00	Macro program call 宏程序调用
G18	02	*ZX* plane selection *ZX* 平面选择	G66	12	Modal macro program call 模态宏程序调用
G19	02	*YZ* plane selection *YZ* 平面选择	* G67	12	Unregister mode macro program call 注销模态宏程序调用
G27	00	Return and check reference point 返回并检查参考点	G73	09	Fixed cycle of deep hole drilling 深孔钻削固定循环
G28	00	Return to reference point 返回参考点	G74	09	Fixed cycle of counter thread tapping 反螺纹攻丝固定循环
G29	00	Return from reference point 从参考点返回	G76	09	Fine boring fixed cycle 精镗固定循环
G30	00	Return to the second reference point 返回第二参考点	* G80	09	Cancel fixed cycle 注销固定循环
* G40	07	Cancel tool radius compensation 注销刀具半径补偿	G81	09	Drilling fixed cycle 钻削固定循环
G41	07	Left tool radius compensation 左侧刀具半径补偿	G84	09	Tapping fixed cycle 攻丝固定循环
G42	07	Right tool radius compensation 右侧刀具半径补偿	* G90	03	Absolute value instruction mode 绝对值指令方式
G43	08	Tool length compensation＋ 刀具长度补偿＋	* G91	03	Incremental value instruction mode 增量值指令方式
G44	08	Tool length compensation－ 刀具长度补偿－	G92	00	Workpiece zero setting 工件零点设定
* G49	08	Cancel tool length compensation 注销刀具长度补偿	G96	02	Constant line speed control 恒线速度控制
G52	00	Set local coordinate system 设置局部坐标系	* G97	02	Constant line speed cancel 恒线速度取消
G53	00	Select machine coordinate system 选择机床坐标系	* G98	05	Feed rate per minute 每分钟进给率
* G54	14	Select No. 1 workpiece coordinate system 选用 1 号工件坐标系	G99	05	Feed rate per revolution 每转进给率

Note: the G instruction with "＊" indicates the status of the G instruction when the power is turned on. The G instructions of group 00 are non modal G instructions, and the others are modal G instructions.

注：带"＊"的 G 指令表示接通电源时，即为该 G 指令的状态。00 组的 G 指令为非模态 G 指令，其他均为模态 G 指令。

Table 9-2 Common M instruction codes for auxiliary functions

表 9-2 常用辅助功能 M 指令代码

Code 代码	Modality 模态	Function description 功能说明	Code 代码	Modality 模态	Function description 功能说明
M00	Modeless 非模态	Program pause 程序暂停	M03	Modal 模态	Spindle forward start 主轴正转启动
M01	Modeless 非模态	Program selection stop 程序选择停止	M04	Modal 模态	Spindle reverse start 主轴反转启动
M02	Modeless 非模态	Program end 程序结束	M05	Modal 模态	Spindle stop 主轴停止转动
M06	Modeless 非模态	Automatic tool changer 自动换刀	M07	Modal 模态	Mist cutting fluid on 雾状切削液打开
M30	Modeless 非模态	End of program and return to the beginning of program 程序结束并返回程序开头	M08	Modal 模态	Cutting fluid on 切削液打开
M98	Modeless 非模态	Call subroutine 调用子程序	M09	Modal 模态	Cutting fluid stop 切削液停止
M99	Modeless 非模态	End of subprogram 子程序结束			

9.2.7 NC Programming

The programming instructions of CNC machine tool vary with the control system, but some commonly used instructions, such as some preparatory functions and auxiliary functions, still conform to ISO standards. Here are some common instructions.

9.2.7.1 Rapid Positioning and Linear Interpolation

(1) Rapid Positioning

The general format of rapid positioning instruction (G00) is

G00 X _ Y _ Z _

where, X _ Y _ Z _ is the coordinate of the end point. When the command is executed, the machine tool moves to the specified position at the maximum moving speed set by itself. The quick positioning command is only used when the tool moves rapidly in the

9.2.7 数控程序编制

数控机床的编程指令随控制系统的不同而不同，但一些常用的指令，如某些准备功能、辅助功能，还是符合 ISO 标准的。下面介绍部分常用的指令。

9.2.7.1 快速定位和直线插补

(1) 快速定位

快速定位指令（G00）的一般格式为

其中，X _ Y _ Z _ 为终点坐标。执行该指令时，机床以自身设定的最大移动速度移向指定位置。快速定位指令仅在刀具非加工状态的快速移动时使用，其功能只是

non-machining state, and its function is only in place rapidly. The motion trajectory varies with the specific CNC system. Generally, it moves to the designated position in a straight line way, but also moves in place along each axis of the polyline in turn. Specilly the feed speed is invalid for the G00 command.

快速到位，其运动轨迹因具体的数控系统不同而不同，一般以直线方式移动到指定位置，也有沿折线每个轴依次移动到位的，且进给速度对G00指令无效。

(2) Linear Interpolation

(2) 直线插补

The general format of the linear interpolation instruction (G01) is

直线插补指令(G01)的一般格式为

G01 X_ Y_ Z_ F_

Where X_ Y_ Z_ is the coordinates of the end point and F is the feed speed. In linear interpolation, the starting point of the straight line is the current position of the tool, which needs not be specified in the program section. The feed speed must be specified in the G01 instruction program segment, or it has been specified in the previous program segment, and this program segment is still valid.

其中，X_Y_Z_为终点坐标，F为进给速度。直线插补加工中，直线的起点是刀具的当前位置，程序段中无须指定。G01指令程序段必须指定进给速度，或者前面程序段已经指定，本程序段续效。

9.2.7.2 Circular Interpolation

9.2.7.2 圆弧插补

The general format of circular interpolation instruction (G02/G03) is

圆弧插补指令(G02/G03)的一般格式为

G17 (G18、G19) G02 (G03) G90 (G91) X_Y_Z_I_J_K_or R_F_

where G17 instruction represents XY plane, G18 instruction represents XZ plane and G19 instruction represents YZ plane. G02 and G03 represent clockwise arc interpolation and counter clockwise circular interpolation respectively. X_ Y_ Z_ represents the position of the end point of the arc. In the G90 absolute input mode, it is the actual coordinate value of the end point of the arc in the part coordinate system. In the G91 incremental input mode, it is the incremental value of the end point of the arc relative to the start point of the arc. I_J_K_ is the incremental value of the center of the circle relative to the starting point of the arc, whether under G90 or G91. In addition, the position of the center of the circle can also be expressed by

其中，G17指令表示XY平面，G18指令表示XZ平面，G19指令表示YZ平面。G02、G03分别表示顺时针圆弧插补、逆时针圆弧插补。X_Y_Z_表示圆弧终点位置，在G90绝对输入方式下为圆弧终点在零件坐标系中的实际坐标值，在G91增量输入方式下为圆弧终点相对于圆弧起点的增量值。I_J_K_为圆心相对于圆弧起点的增量值，不论是在G90下还是在G91下都是如此。另外，圆心的位置也可以用圆弧的半径R表示。当圆弧所对应的圆心角超过

the radius R of the arc. When the center angle corresponding to the arc exceeds 180°, the radius R is represented by a negative value. When it is exactly 180°, it can be positive or negative. But R and I _ J _ K _ can not be mixed In the program. In addition, R cannot be used in the whole circle programming, but I _ J _ K _ can only be used.

180°时，半径 R 用负值表示；正好为 180°时，正负均可。但用 R 时不能用 I _ J _ K _ ，程序中 R 与 I _ J _ K _ 二者不能混用。另外，整圆编程时不能使用 R，只能用 I _ J _ K _ 。

9.2.7.3 Tool Radius Compensation

9.2.7.3 刀具半径补偿

The general format of the tool radius compensation command (G40, G41, G42) is

刀具半径补偿指令（G40，G41，G42）的一般格式为

G00/G01 G41/G42 D X _ Y _ F
G00/G01 G40 X _ Y _

where G40 means to cancel the tool radius compensation; G41 represents left compensation of tool diameter (compensation on the left in the direction of tool advance); G42 represents right compensation of tool diameter (compensation on the right in the direction of tool advance), as shown in Fig. 9-16.

其中，G40 表示取消刀具半径补偿；G41 表示左刀补偿（在刀具前进方向左侧补偿）；G42 表示右刀补偿（在刀具前进方向右侧补偿），如图 9-16 所示。

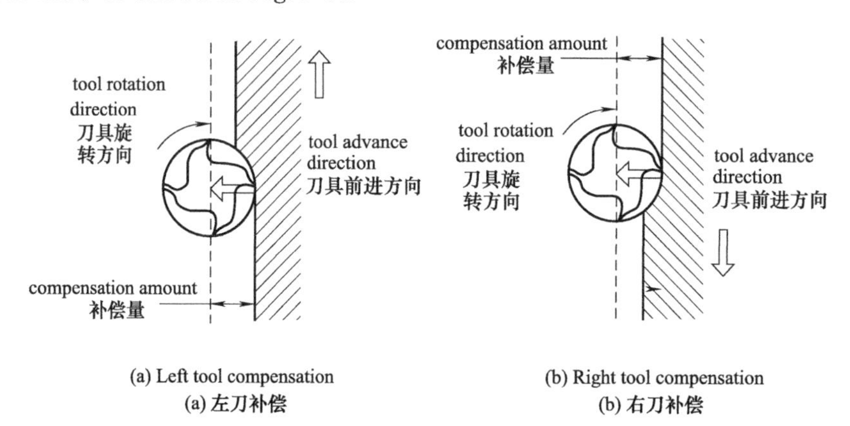

Fig. 9-16 Tool radius compensation direction
图 9-16 刀具半径补偿方向

D in the first command is the tool radius compensation address, in which the compensation amount of tool radius is stored. X _ Y _ is the starting position from the non tool compensation state to the tool radius compensation state. In the second command, X _ Y _ is the end position from the non-tool compensation state to the tool compensation state, where X _ Y _ is the position of tool center.

第一个指令中的 D 为刀具半径补偿地址，地址中存放的是刀具半径的补偿量；X _ Y _ 为由非刀补状态进入刀具半径补偿状态的起始位置。第二个指令中的 X _ Y _ 为由刀补状态过渡到非刀补状态的终点位置，这里的 X _ Y _ 即为刀具中心的位置。

9.2.7.4 Subprogram Instruction Format

9.2.7.4 子程序指令格式

In order to simplify the programming, when a part has the same machining content, the method of adjusting the subprogram is often used. The program that calls the subprogram is called the main program. The number of the subprogram is basically the same as that of general program, except that M99 represents the end of the subprogram and returns to the main program calling the subprogram. Programming format of calling subprogram is

编程时，为了简化程序的编制，当一个零件上有相同的加工内容时，常用调子程序的方法进行编程。调用子程序的程序称为主程序。子程序的编号与一般程序基本相同，只是以M99表示子程序结束，并返回到调用子程序的主程序中。调用子程序的编程格式为

```
M98 P        //Number of calls to program number L 程序号 L 调用次数
O10          //Subroutine program number 子程序程序号
N01          //Subroutine body 子程序体
N0n M99      //Subprogram ends and returns to main program 子程序结束并返回主程序
```

【Example 9-1】 Programming the CNC turning program of the parts shown in Fig. 9-17.

【例 9-1】 编制图 9-17 所示零件的数控车削程序。

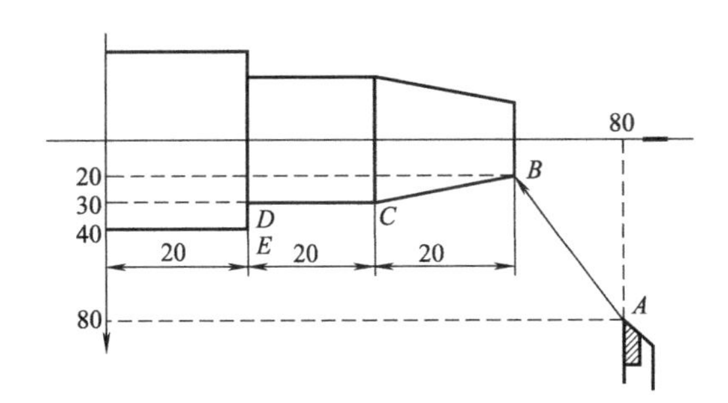

Fig. 9-17 Figure for example 9-1

图 9-17 例 9-1 附图

```
O0001
N1 M03 S640          //Spindle rotation（主轴正转）
N2 G90               //Select absolute increment（选绝对增量）
N3 G00X20Z60         //Fast forward A →B（快进 A →B）
N4 G01 X30 Z40 F100  //Working in B →C（工进 B →C）
N5 G01X30Z20         //C →D
N6 G01X40Z20         //D →E
N7 M02               //Program end（程序结束）
```

【Example 9-2】 Programming the finishing program of the inner contour of the part as shown in Fig. 9-18. The tool radius is 8mm, the programming origin is built on the upper surface of the part center, and the left tool is used for patching.

【例 9-2】 编写图 9-18 所示零件内轮廓的精加工程序，刀具半径为 8mm，编程原点建在零件中心上表面，用左刀补加工。

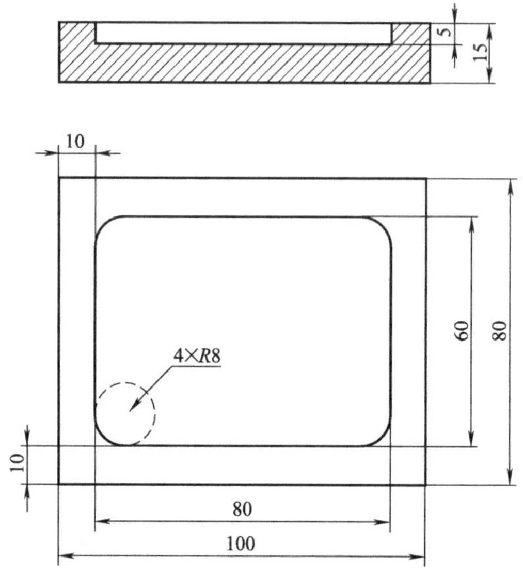

Fig. 9-18 Figure for example 9-2

图 9-18 例 9-2 附图

```
O0002
N01 G90 G92 X0 Y0 Z100        //Set workpiece coordinate system 设定工件坐标系
N02 T01 M06                   //Automatic tool changer 自动换刀
N03 M03 S500                  //Spindle rotation 主轴正转
N04 G00 G43 H01 Z5            //Tool length compensation 刀具长度补偿
N05 G01 Z-5 F100
N06 G41 G01 X40 Y0 D01 F200   //Tool left radius compensation 刀具左侧半径补偿
N07 Y30                       //Interpolation along the inner contour of workpiece
                                沿工件内轮廓进行插补加工
N08 X-40
N09 Y-30
N10 X40
N11 Y2
N12 G40 G01 X0 Y0             //Write off tool left radius 刀具左侧半径注销
N13 G49 G00 Z100              //Cancel tool length compensation 刀具长度补偿注销
N14 M30                       //Program end 程序结束
```

9.3 CNC Interpolation Principle

9.3.1 CNC Interpolation Overview

For CNC, various kinds of contour machining are realized via interpolation calculation. The task of interpolation calculation is to calculate a limited number of coordinate points between the starting point and the end point of the contour line. Then the tool moves along these coordinate points to approximate the theoretical contour.

Interpolation methods can be divided into two categories: pulse incremental interpolation and data sampling interpolation.

Pulse incremental interpolation is an interpolation method to control the output law of a single pulse. Each time a pulse is output, the moving part must move a certain distance correspondingly. This distance is called the pulse equivalent. Therefore, pulse incremental interpolation is also called stroke scalar interpolation. Such as point-by-point comparison method, digital integration method. Depending on the processing accuracy, pulse equivalent can be selected from 0.01 to 0.001mm. The moving speed of moving parts is related to pulse equivalent and pulse output frequency, which is usually tens of thousands of Hertz. Therefore, when the pulse equivalent is 0.001mm, the maximum moving speed is only about 2 m/min. Pulse incremental interpolation is usually used in stepping motor control system.

The data sampling interpolation method (also known as digital incremental interpolation) is a method which calculating the increment value of each coordinate direction (Δx, Δy, Δz), the coordinate position of the tool and other required values within the specified time (also called as interpolation cycle).

9.3 数控插补原理

9.3.1 数控插补概述

对 CNC 来说，各种轮廓加工都是通过插补计算实现的，插补计算的任务就是在轮廓线的起点到终点之间计算出有限个坐标点，刀具沿着这些坐标点移动，来逼近理论轮廓。

插补方法可分两大类：脉冲增量插补和数据采样插补。

脉冲增量插补是控制单个脉冲输出规律的插补方法。每输出一个脉冲，移动部件都要相应的移动一定距离，这个距离称为脉冲当量，因此，脉冲增量插补也叫做行程标量插补，如逐点比较法、数字积分法。根据加工精度的不同，脉冲当量可取 0.01 ～ 0.001mm。移动部件的移动速度与脉冲当量和脉冲输出频率有关，脉冲输出频率通常为几万赫兹。因此，当脉冲当量为 0.001mm 时，最高移动速度仅有 2m/min 左右。脉冲增量插补通常用于步进电机控制系统。

数据采样插补法（也称数字增量插补法）是在规定的时间（称作插补周期）内，计算各坐标方向的增量值（Δx，Δy，Δz）、刀具所在的坐标位置及其他一些需要的值。这些数据严格的限制在一个

These data are strictly limited in one interpolation cycle (such as 2ms) and sent to the servo system, which controls the movement of moving parts. The moving part must also finish the stroke given by interpolation calculation in the next interpolation time, so data sampling interpolation is also called time scalar interpolation.

插补周期内（如 2ms）计算完毕，送给伺服系统，再由伺服系统控制移动部件运动。移动部件也必须在下一个插补时间内走完插补计算给出的行程，因此数据采样插补也称作时间标量插补。

Because the data sampling interpolation method uses numerical values to control the movement of the machine tool. Therefore, the movement speed of each coordinate direction of the machine tool is related to the numerical value given by the interpolation operation and the interpolation time. Depending on the operating speed and processing accuracy of the computer, the interpolation cycle of the CNC system is usually between 1ms and 10ms. Thus, the feed speed of modern CNC machine tool has exceeded 30m/min and some of them have reached 120m/min. The data sampling interpolation method is suitable for the closed-loop or semi-closed-loop control system of AC servo motor.

由于数据采样插补法是用数值量控制机床运动，因此，机床各坐标方向的运动速度与插补运算给出的数值和插补时间有关。根据计算机运行速度和加工精度不同，数控系统的插补周期通常在 1～10ms 之间。现代数控机床的进给速度已超过 30m/min，有些已达到 120 m/min。数据采样插补法适用于交流伺服电机的闭环或半闭环控制系统。

Take the point-by-point comparison linear interpolation as an example to briefly introduce the realization principle of CNC interpolation.

下面以逐点比较法直线插补为例简单介绍数控插补实现原理。

9.3.2 Linear Interpolation with Point-by-point Comparison Method

9.3.2 逐点比较法直线插补

The linear interpolation with point-by-point comparison method is also called algebraic operation method and drunken step method. It is one of the earliest interpolation algorithms and belongs to pulse incremental interpolation. The principle is: in the process of control processing, CNC system can calculate and discriminate the deviation between the motion trajectory of the tool and the given trajectory point-by-point, and control the feed axis to approach the given contour direction according to the deviation, so that the processing contour can approach the given contour curve. The

逐点比较插补法又称代数运算法、醉步法，它是一种最早的插补算法，属于脉冲增量插补。其原理是：在控制加工过程中，CNC 系统能逐点计算和判别刀具的运动轨迹与给定轨迹的偏差，并根据偏差控制进给轴向给定轮廓方向靠近，使加工轮廓逼近给定轮廓曲线。逐点比较法以折线来逼近直线或圆弧曲线，它与给定的直线或圆弧之间的最大误差不超过一

point-by-point comparison method approximates a straight line or arc curve with a broken line, and the maximum error between it and the given straight line or arc does not exceed one pulse equivalent. Therefore, as long as the pulse equivalent (i. e., the distance for the coordinate axis to feed one step) is small enough, the requirements of machining accuracy can be met.

个脉冲当量，因此只要将脉冲当量（即坐标轴进给一步的距离）取得足够小，就可满足加工精度的要求。

Suppose that the straight line OA in the first quadrant as shown in Fig. 9-19 is processed. The starting point of the line is the origin of the coordinate and the coordinate of the end point $A(x_e, y_e)$ is known, let $P(x_i, y_i)$ is any of a processing point (moving point). If the point P is exactly on a straight line, then $y_i/x_i = y_e/x_e$ holds, i. e., $y_i x_e - y_e x_i = 0$. If the point P is above the line (point P' in the figure), then $y_i/x_i > y_e/x_e$ holds, i. e., $y_i x_e - y_e x_i > 0$. If the point P is below the straight line (point P'' in the figure), then $y_i/x_i < y_e/x_e$ holds, i. e., $y_i x_e - y_e x_i < 0$.

假设加工如图 9-19 所示第一象限的直线 OA。直线的起点为坐标原点，直线的终点坐标 $A(x_e, y_e)$ 为已知，设 $P(x_i, y_i)$ 为任意一个加工点（动点）。若 P 点正好在直线上，则 $y_i/x_i = y_e/x_e$ 成立，即 $y_i x_e - y_e x_i = 0$；若 P 点在直线的上方（图中 P' 点），则 $y_i/x_i > y_e/x_e$ 成立，即 $y_i x_e - y_e x_i > 0$；若 P 点在直线的下方（图中 P'' 点），则 $y_i/x_i < y_e/x_e$ 成立，即 $y_i x_e - y_e x_i < 0$。

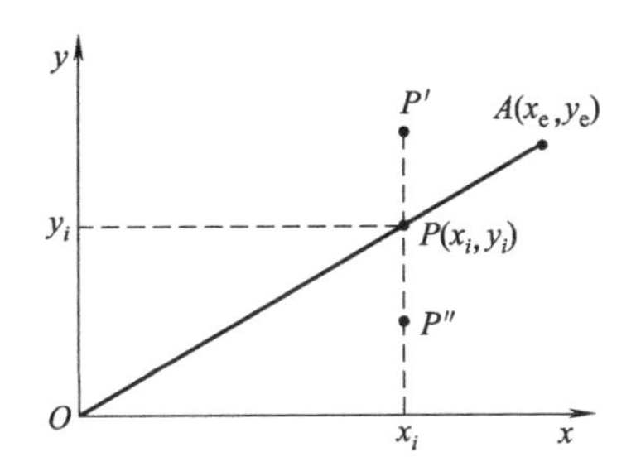

Fig. 9-19 Linear interpolation with point-by-point comparison method

图 9-19 逐点比较法直线插补

Define the linear interpolation deviation judgment formula as $F_i = y_i x_e - y_e x_i$. If $F_i = 0$, it indicates that the point P is on the straight line OA. If $F_i > 0$, it indicates that the point P is above the straight line OA. If $F_i < 0$, it indicates that the point P is below the straight line OA.

定义直线插补的偏差判别式为 $F_i = y_i x_e - y_e x_i$。若 $F_i = 0$，则表明 P 点在直线 OA 上；若 $F_i > 0$，则表明 P 点在直线 OA 的上方；若 $F_i < 0$，则表明 P 点在直线 OA 的下方。

As seen from the Fig. 9-19, when $F_i \geqslant 0$, the tool should further feed one step in the $+x$ direction to approximate the given straight line and the coordinate value at this time is

从图 9-19 中可以看出：当 $F_i \geqslant 0$ 时，刀具应向 $+x$ 方向进给一步，以逼近给定直线，此时的坐标值为

$$\begin{cases}x_{i+1}=x_i+1\\y_{i+1}=y_i\end{cases} \tag{9-1}$$

The deviation of the new processing point is

则新加工点的偏差为

$$F_{i+1}=y_{i+1}x_e-y_ex_{i+1}=y_ix_e-y_e(x_i+1) \tag{9-2}$$

When $F_i<0$, the tool should feed one step in the $+y$ direction and the coordinate value at this time is

当若 $F_i<0$ 时，刀具应向 $+y$ 方向进给一步，此时的坐标值为

$$\begin{cases}x_{i+1}=x_i\\y_{i+1}=y_i+1\end{cases} \tag{9-3}$$

The deviation of the new processing point is

则新加工点的偏差为

$$F_{i+1}=y_{i+1}x_e-y_ex_{i+1}=(y_i+1)x_e-y_ex_e=y_ix_e-y_ex_i+x_e=F_i+x_e \tag{9-4}$$

After feeding one step, the deviation of the new processing point can be calculated from the processing deviation of the previous point and the coordinate of the end point (x_e, y_e). Then determine the next step according to the sign of the deviation discriminant of the new processing point. Continue in this way until the coordinate values in the two directions are equal to the coordinate of the end point (x_e, y_e). Then the arrival signal of the end point will be sent and the interpolation of the linear segment ends.

进给一步后，由前一点的加工偏差和终点坐标（x_e，y_e）可计算出新加工点的偏差，再根据新加工点偏差判别式的符号决定下一步的走向。如此下去，直到两个方向的坐标值与终点坐标（x_e，y_e）相等，发出终点到达信号，该直线段插补结束。

In summary, in the point-by-point comparison method, the following four items must be completed for each step of the tool feed.

综上所述，逐点比较法中刀具每进给一步都要完成以下 4 项内容：

• Deviation sign discrimination. Determine whether the processing point is outside or inside the outline of the specified part, that is, judge whether $F\geqslant0$.

•偏差符号判别。确定加工点是在规定零件轮廓线外侧还是内侧，即判断是否 $F\geqslant0$。

• Coordinate feed. According to the deviation, control the x-coordinate or the y-coordinate to feed one step to make the processing point move closer to the contour of the part to reduce the deviation. When $F\geqslant0$, feed one step in the $+x$ direction. While $F<0$, feed one step in the $+y$ direction.

•坐标进给。根据偏差情况，控制 x 坐标或 y 坐标进给一步，使加工点向零件轮廓线靠拢，以缩小偏差。当 $F\geqslant0$ 时，向 $+x$ 方向进给一步；当 $F<0$ 时，向 $+y$ 方向进给一步。

• Calculate the new deviation. After feeding one step, the deviation between the new processing point and the

•新偏差计算。进给一步后，计算新加工点与零件轮廓的偏差，作

contour of the part is calculated as the basis for the deviation judgment in the next step. The formula is the Eq. (9-2) or Eq. (9-4).

为下一步偏差判别的依据。计算公式为式(9-2) 或式(9-4)。

• End point discrimination. There are two ways to determine the end point. One is to calculate the total number of steps to be fed in the x and y directions, i. e., $\sum N=(|x_e|-x_o)+(|y_e|-y_o)=|x_e|+|y_e|$. This means that for each step in the x or y direction, $\sum N$ minus 1 is calculated. When $\sum N$ is reduced to zero, the end point is reached and the interpolation is stopped. Another method is to obtain the number of steps to be fed in the x-coordinate and y-coordinate respectively, i. e., the values of $|x_e|$ and $|y_e|$. When feeding one step in the x direction, N_x-1. When feeding one step in the y direction, N_y-1. When both N_x and N_y are zero, the end point is reached and the interpolation is stopped.

•终点判别。判别终点的方法有两种：一是计算出 x 和 y 方向坐标所要进给的总步数，即$\sum N=(|x_e|-x_o)+(|y_e|-y_o)=|x_e|+|y_e|$，每向 x 或 y 方向进给一步，均进行$\sum N$ 减 1 计算，当$\sum N$ 减至零时即到终点，停止插补。另一种方法是分别求出 x 坐标和 y 坐标应进给的步数，即 $|x_e|$ 和 $|y_e|$ 的值，当沿 x 方向进给一步时，N_x-1；当沿 y 方向进给一步时，N_y-1；当 N_x 和 N_y 都为零时，达到终点，停止插补。

【Example 9-3】 Suppose the linear segment OA is interpolated in the first quadrant, the starting point is the coordinate origin O (0, 0) and the end point is A (8, 6). Try to use the point-by-point comparison method to interpolate and draw the interpolation trajectory.

【例 9-3】 设在第一象限插补直线段 OA，起点为坐标原点 O (0, 0)，终点为 A (8, 6)。试用逐点比较法进行插补，并画出插补轨迹。

Solution Use the first end point judgment method to interpolate this straight line. The total steps of the tool should feed along the x and y axes is $\sum N=|x_e|+|y_e|=8+6=14$. The interpolation operation process is shown in Table 9-3.

解 用第一种终点判别法插补这段直线，刀具沿 x、y 轴应进给的总步数为$\sum N=|x_e|+|y_e|=8+6=14$。插补运算过程如表 9-3 所示。

Table 9-3 Interpolation calculation process with the point-by-point comparison method

表 9-3 逐点比较插补运算过程

Analyzing deviation 偏差判断	Feed direction 进给方向	The new deviation calculation 新偏差计算	End judgment 终点判别
$F_0=0$	$+x$	$F_1=F_0-y_e=0-6=-6$	$\sum N=14-1=13$
$F_1=-6<0$	$+y$	$F_2=F_1+x_e=-6+8=2$	$\sum N=13-1=12$
$F_2=2>0$	$+x$	$F_3=F_2-y_e=2-6=-4$	$\sum N=12-1=11$
$F_3=-4<0$	$+y$	$F_4=F_3+x_e=-4+8=4$	$\sum N=11-1=10$
$F_4=4>0$	$+x$	$F_5=F_4-y_e=4-6=-2$	$\sum N=10-1=9$
$F_5=-2<0$	$+y$	$F_6=F_5+x_e=-2+8=6$	$\sum N=9-1=8$
$F_6=6>0$	$+x$	$F_7=F_6-y_e=6-6=0$	$\sum N=8-1=7$

续表

Analyzing deviation 偏差判断	Feed direction 进给方向	The new deviation calculation 新偏差计算	End judgment 终点判别
$F_7=0$	$+x$	$F_8=F_7-y_e=0-6=-6$	$\sum N=7-1=6$
$F_8=-6<0$	$+y$	$F_9=F_8+x_e=-6+8=2$	$\sum N=6-1=5$
$F_9=2>0$	$+x$	$F_{10}=F_9-y_e=2-6=-4$	$\sum N=5-1=4$
$F_{10}=-4<0$	$+y$	$F_{11}=F_{10}+x_e=-4+8=4$	$\sum N=4-1=3$
$F_{11}=4>0$	$+x$	$F_{12}=F_{11}-y_e=4-6=-2$	$\sum N=3-1=2$
$F_{12}=-2<0$	$+y$	$F_{13}=F_{12}+x_e=-2+8=6$	$\sum N=2-1=1$
$F_{13}=6>0$	$+x$	$F_{14}=F_{13}-y_e=6-6=0$	$\sum N=1-1=0$

The linear interpolation with point-by-point comparison method can be realized by hardware or software. When implemented by hardware, two coordinate registers, deviation registers, adders, end point discriminators and so on are used to form a logic circuit to realize the linear interpolation with point-by-point comparison method. The achieved program flow of the interpolation via software is shown in Fig. 9-20 and the interpolation trajectory is shown in Fig. 9-21.

逐点比较法直线插补可以用硬件实现，也可以用软件实现。用硬件实现时，采用两个坐标寄存器、偏差寄存器、加法器、终点判别器等组成逻辑电路，即可实现逐点比较法的直线插补。用软件实现插补的程序流程如图 9-20 所示，插补轨迹如图 9-21 所示。

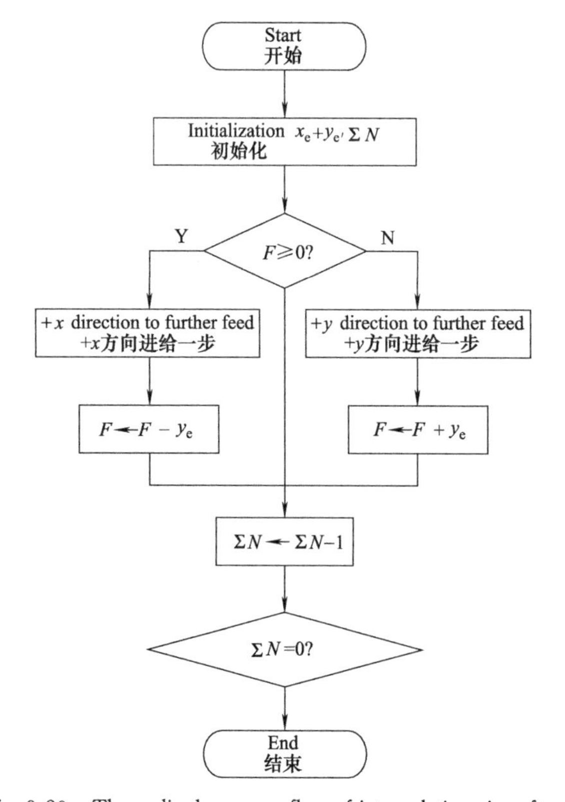

Fig. 9-20 The realized program flow of interpolation via software

图 9-20 软件实现插补的程序流程

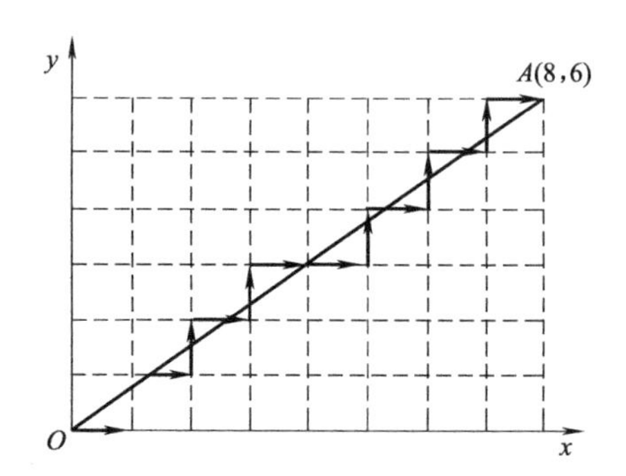

Fig. 9-21 Interpolation trajectory

图 9-21 插补轨迹

9.4 Key Performance Indexes of CNC Machine Tool

9.4.1 Main Technical Specifications

The technical specifications of CNC machine tool lathe mainly include the maximum turning diameter between the bed and the tool rest, the maximum turning length and the maximum turning diameter. The technical specifications of CNC milling machines mainly include the specifications and dimensions of the worktable, the T-slot of the worktable, and the stroke of the worktable.

9.4.2 Motion Index

(1) Spindle Speed

The spindle of the CNC machine tool is driven by an AC motor and is supported by high-speed precision bearings to ensure that the spindle has a wide range of speed adjustment, higher rotation accuracy, and higher stiffness and shock resistance. The spindle speed of modern CNC generally reaches 5000～10000rpm or even higher, which is extremely beneficial to improve the processing quality and small hole processing.

(2) Feed Rate

Feed rate is the main factor that affects processing quality, production efficiency and tool life. At present, the feed speed of CNC can reach 10～30m/min, where the maximum fast forward speed is the maximum speed of movement when not processing, and the maximum feed speed is the maximum speed during processing.

9.4 数控机床的主要性能指标

9.4.1 主要技术规格

数控车床主要技术参数有床身与刀架最大回转直径、最大车削长度、最大车削直径等；数控铣床主要技术参数有工作台、工作台T形槽、工作台行程等规格尺寸。

9.4.2 运动指标

(1) 主轴转速

数控机床主轴采用交流电动机驱动，选用高速精密轴承支承，保证主轴具有较宽的调速范围和较高的回转精度，以及较高的刚度和抗震性。现代数控机床的主轴普遍达到5000～10000r/min甚至更高的转速，这对提高加工质量和小孔加工极为有利。

(2) 进给速度

进给速度是影响加工质量、生产效率、刀具寿命的主要因素。目前，数控机床的进给速度可达到10～30m/min，其中最大快进速度为不加工时移动的最大速度，最大进给速度为加工时的最大速度。

9.4.3 Precision Index

(1) Pulse Equivalent (resolution)

Pulse equivalent is the main factor affecting the machining accuracy and surface quality of CNC, which is an important precision index of CNCs. The pulse equivalent of ordinary CNC is 0.001mm. The pulse equivalent of simple CNC is generally 0.01mm. And the pulse equivalent of precision or ultra-precision CNC is 0.0001mm.

(2) Positioning accuracy

Positioning accuracy refers to the accuracy of the actual position that can be achieved by moving parts such as the workbenches of CNC. The difference between the actual position and the command position is the positioning error. The factors that cause positioning errors come from servo system, detection system, feed system and the geometric error of moving parts. The positioning error directly affects the dimensional accuracy of the parts being processed. Generally, the positioning accuracy of CNC is ±0.01mm.

(3) Repositioning Accuracy

Repositioning accuracy refers to the consistency degree of obtained results when the same operation is repeated by the same operation method under the same condition. Repositioning accuracy is generally a contingency error with a normal distribution, which will affect the consistency of parts processed in batches. In general, the repositioning accuracy of CNC is ± 0.005mm.

9.4.3 精度指标

（1）脉冲当量（分辨率）

脉冲当量是影响数控机床加工精度和表面质量的主要因素，是数控机床的重要精度指标。普通数控机床的脉冲当量为0.001mm，简易数控机床的脉冲当量一般为0.01mm，精密或超精密数控机床的脉冲当量为0.0001mm。

（2）定位精度

定位精度是指数控机床工作台等移动部件所达到的实际位置的精度。实际位置与指令位置的差值为定位误差。引起定位误差的因素来源于伺服系统、检测系统、进给系统，以及运动部件的几何误差。定位误差直接影响零件加工的尺寸精度，一般数控机床的定位精度为±0.01mm。

（3）重复定位精度

重复定位精度是指在相同的条件下，采用相同的操作方法，重复进行同一动作时，得到结果的一致程度。重复定位精度一般是呈正态分布的偶然性误差，它会影响批量加工零件的一致性。一般数控机床的重复定位精度为±0.005mm。

9.4.4 Tool System

CNC machine tool includes the number of tool holder, tool hole diameter, tool bar size, tool change time and so on.

The tool storage capacity and tool change time directly affect the productivity of machining center. Generally, the tool capacity in small and medium size machining centers is 16 to 60, and that in large machining centers is more than 100. The tool change time is generally 1 to 20 s.

9.4.4 刀具系统

数控机床包括刀架位数、刀具孔直径、刀杆尺寸、换刀时间等各项内容。

刀库容量与换刀时间直接影响加工中心的生产率。通常中小型加工中心的刀库容量为 16～60 把，大型加工中心可达 100 把以上。换刀时间一般为 1～20s。

9.4.5 Other Indexes

In addition to the above performance indexes, there are electrical indexes such as the specifications, models and power of the main motor and servo motor, the indexes of the cooling system, the dimension and weight of CNC and so on.

9.4.5 其他指标

除上述性能指标外，还有主电动机、伺服电动机规格型号和功率等电气指标，冷却系统指标，数控机床外形尺寸、重量等。

9.5 Development of the NC Technology

The development of CNC machine tool began in the United States. In 1952, Parsons and the Massachusetts Institute of Technology successfully developed the first three-axis CNC milling in the world. In 1955, CNC machine tool began to enter the practical stage, which have been mainly used for complex surface processing in the aviation industry.

In the early 1970s, with the development of computer technology, many functions of NC can be realized by special programs. And these special programs can be stored in the memory of a small computer, which is the so-called soft wiring NC, that is the computer control system CNC.

9.5 数控技术发展

数控机床的研制始于美国，1952 年帕森斯公司和麻省理工学院合作，成功研制了世界第一台三坐标数控铣床，1955 年进入实用阶段，主要用于航空工业的复杂曲面加工。

20 世纪 70 年代初，计算机技术的发展使得数控的许多功能可以用编制的专用程序来实现，而这些专用程序可以存储在小型计算机的存储器中，这就是所谓的软接线数控，即计算机控制系统 CNC。

With the further rapid development of NC technology in recent years, the following representative CNC systems have appeared.

随着近年来数控技术的进一步快速发展，出现了下列具有代表性的数控系统。

(1) Distributed Numerical Control

(1) 计算机直接控制系统

Distributed Numerical Control (DNC) is also known as the group control. It's characteristics are: Using computer to strengthen the management of the production process so that the programming, production preparation and planning arrangements are coordinated with the machine tool work, thereby improving the efficiency of each CNC.

计算机直接控制系统又称群控，其特点是：使用计算机加强对生产过程的管理，使程序的编制、生产的准备与计划安排等工作和机床工作协调一致，以提高各个数控机床的使用效率。

(2) Adaptive Control Machine Tool Machine Tool

(2) 自适应控制机床

Generally, CNC machine tool is processed according to pre-programmed programs. However, many parameters during programming can only be determined by empirical data and it is impossible to accurately consider all their changes, such as the unevenness of the blank, the change of the material of the tool and part, the wear of the tool, the deformation of the part, the difference in thermal conductivity and so on. These changes directly or indirectly affect the quality of processing, making processing impossible to proceed in the best state. If the control system can timely measure the parameters of various machining states and feed it back to the machine tool for correction, the cutting process can be in the best state at any time. The so-called best state refers to the highest productivity, the lowest processing cost, the best processing quality and so on. As the CNC system has its own computer, this kind of adaptive control machine tool can be made by adding corresponding detection elements, control circuits and relevant software.

一般数控机床是按预先编好的程序进行加工的，但编程时的许多参数只能参照经验数据来决定，不可能准确地考虑到它们的一切变化，如毛坯的不均匀、刀具与零件材质的变化、刀具的磨损、零件的变形、热传导性的差别等，这些变化直接或间接地影响加工质量，使加工不能在最佳状态下进行。如果控制系统能及时测量实际加工中的各种加工状态的参数并反馈给机床进行修正，则可使加工过程随时都处在最佳状态。所谓最佳状态，指的是最高生产率、最低加工成本、最好的加工质量等。由于 CNC 系统自身带有计算机，只要加上相应的检测元件、控制线路和有关软件，就可以制造出这种自适应控制机床。

(3) Flexible Manufacturing System

Flexible manufacturing system (FMS) is developed on the basis of flexible manufacturing cell (FMC). FMC is the smallest unit that can automatically process different parts in the same group of parts (including the transportation and exchange of parts in the cell). It can be used not only as an independent processing equipment, but also as a basic component module of a larger and more complex flexible manufacturing system or flexible automatic line. FMS is composed of processing system (which is composed of a group of CNC machine tools and other automatic process equipment, such as washing machine, finished product testing machine, paint spraying machine, etc.), intelligent robot, automatic conveying system and automatic warehouse (Fig. 9-22). This system can process a group of parts with different processes and different processing tempo in any order. The process flow can be adjusted with different parts. The entire production process is scheduled by the central computer. And several computers are used for station control. Each manufacturing unit is relatively independent and can balance the utilization of resources in a timely manner.

(3) 柔性制造系统

柔性制造系统是在柔性制造单元的基础上研制和发展起来的。柔性制造单元是指能连续地对同一组零件内不同的零件进行自动化加工（包括零件在单元内部的运输和交换）的最小单元。它既可以作为独立使用的加工设备，又可以作为更大、更复杂的柔性制造系统或柔性自动线的基本组成模块。柔性制造系统是由加工系统（由一组数控机床和其他自动化工艺设备，如清洗机、成品试验机、喷漆机等组成）、智能机器人、全自动输送系统及自动化仓库组成（图 9-22）。这种系统可按任意顺序加工一组不同工序与不同加工速度的零件，工艺流程可随零件不同而调整，生产过程由中央计算机进行生产程序的调度，由若干台计算机进行工位控制。其中各个制造单元相对独立，能适时地平衡资源的利用。

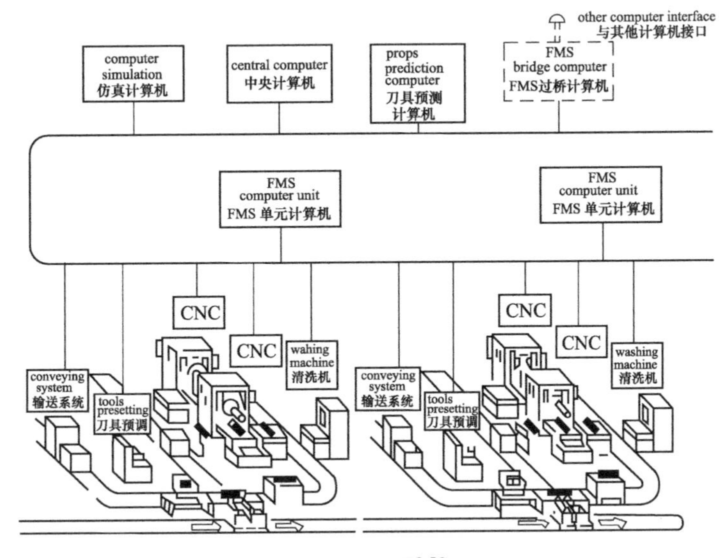

Fig. 9-22 FMS

图 9-22 柔性制造系统

(4) Computer Integrated Manufacturing Systems

With the rapid popularization of intelligent manufacturing, people have further advanced an integrated manufacturing system which the entire process including production planning, design, technology, processing, assembly, inspection, and sales is controlled by a computer. It has automatic information flow, material flow and energy flow controlled by computer. It can control the whole process from product conception and design to final assembly and inspection, so as to realize the great goal of digital workshop and intelligent factory.

(4) 计算机集成制造系统

随着智能制造的快速普及，人们进一步提高了制造生产计划、设计、工艺、加工、装配、检验、销售等全过程均由计算机控制的集成制造系统的能力。这种系统具有计算机控制的自动化信息流、物质流和能量流，可以控制产品构思、设计直到最终装配、检验的全过程，从而实现数字化车间及智能工厂这一伟大的目标。

References

参 考 文 献

[1] 王树青.自动化专业英语.3版.北京：化学工业出版社，2010.

[2] 李国厚，赵欣.自动化专业英语.2版.北京：北京大学出版社，2015.

[3] 黄星，刘治满.电气自动化专业英语.3版.北京：人民邮电出版社，2018.

[4] 辛宗生，魏国丰.自动化制造系统.北京：北京大学出版社，2012.

[5] 全燕鸣.机械制造自动化.广州：华南理工大学出版社，2008.

[6] 江桂云.机械电气控制及自动化.北京：机械工业出版社，2014.

[7] 凌跃胜，等.电气工程及其自动化专业英语教程.北京：中国电力出版社，2019.

[8] 戴文进.电气工程及其自动化专业英语.2版.北京：电子工业出版社，2018.

[9] 杨宜民，等.自动化科学与技术概论.北京：清华大学出版社，2015.

[10] Yusuf A.制造自动化：金属切削力学、机床振动和CNC设计（英文版）.2版.北京：电子工业出版社，2019.

[11] 刘岩川，张艳.自动化与检测技术.大连：大连理工大学出版社，2014.

[12] 郭明良，等.电气控制与西门子PLC应用技术.北京：化学工业出版社，2018.

[13] 胡海清，刘雪雪.PLC与自动生产线技术.北京：北京理工大学出版社，2010.

[14] Peter C. Robotics，Vision and Control：Fundamental Algorithms In MATLAB：Second Edition. Springer.

[15] Peter C.机器人学、机器视觉与控制：MATLAB算法基础.刘荣，等译.北京：电子工业出版社，2016.

[16] Luc J.机器人自动化：建模、仿真与控制.黄心汉，彭刚，译.北京：机械工业出版社，2017.

[17] 吴振顺.气压传动与控制.哈尔滨：哈尔滨工业大学出版社，2009.

[18] 吴晓明，等.现代气动元件与系统.北京：化学工业出版社，2014.

[19] 董林福，赵艳春，刘希敏，等.气动元件与系统识图.北京：化学工业出版社，2009.

[20] 王华.控制工程基础.北京：北京航空航天大学出版社，2009.

[21] 黄筱调，丁文政，洪荣晶.机床数控化改造理论、方法及应用.北京：科学出版社，2012.